According to the official historian Brigadier-General James Edmonds: 'In every respect the Expeditionary Force of 1914 was incomparably the best trained, best organized, and best equipped British Army which ever went forth to war'. There has been considerable debate over the extent to which Edmonds' claim was justified, and to which the British Army had learnt the lessons of recent events (above all, its chastening experiences in South Africa). Conventional wisdom has it that the British Army in 1914 was utterly unprepared for the development of trench warfare from October 1914 onwards, and that it took many lives and a costly 'learning curve' for the British to come to terms with the new conditions of warfare. Given that war was expected in the decade before August 1914 – and that a great deal of time and money was spent preparing for that war – it seems obvious to ask why the British Army was not better prepared for the war when it came. This raises important issues about how armies learn from their experiences and how they prepare for the unknowable – namely, a war – without employing bullets and shells. How realistic and useful were the exercises and manoeuvres the British Army used in the period between the end of the Boer War in 1902 and the outbreak of war in August 1914? The approach of most historians has been either to ignore them, or to dismiss them as a waste of time and money. The manoeuvres carried out between 1902 and 1913 featured large forces – sometimes as many as 45,000 men and 12,000 horses – as well as guns, trucks, trains and the first sizable force of military aircraft ever employed in Britain. Many of the names later familiar from the Western Front were involved – Haig, French, Rawlinson and Allenby – as well as a great many of the troops who would cross to France with the British Expeditionary Force (BEF) in August 1914. Their efforts were witnessed by large crowds, as well as politicians, representatives of foreign armies and journalists (some of them 'embedded' with army units); there was comprehensive and opinionated coverage in the newspapers of the time. What lessons were learnt, what value did these manoeuvres have and how do they relate to the events of the war – especially, its opening months? How does the British experience compare with those of the continental armies, who also made extensive use of manoeuvres in this period?

Simon Batten studied Modern History at Jesus College, Oxford, where the experience of studying under Sir Michael Howard inspired him to pursue his interest in military history. He has taught History at Bloxham School since 1985, is the school archivist and also coaches rugby. Simon has conducted numerous battlefield tours and has written articles on British commanders in the First World War, as well as lecturing on the subject. His twin interests in sports coaching and military history have led to a fascination with the question of how armies learn from their experiences and how they prepare for war. He is the author of *A Shining Light* (2010).

# FUTILE EXERCISE?

## *The British Army's Preparations for War 1902-1914*

---

Simon Batten

Helion & Company Limited
Unit 8 Amherst Business Centre
Budbrooke Road
Warwick
CV34 5WE
England
Tel. 01926 499 619
Email: info@helion.co.uk
Website: www.helion.co.uk
Twitter: @helionbooks
Visit our blog https://helionbooks.wordpress.com/

Published by Helion & Company 2018. Reprinted in paperback 2024
Designed and typeset by Mach 3 Solutions (www.mach3solutions.co.uk)
Cover designed by Paul Hewitt, Battlefield Design (www.battlefield-design.co.uk)



ISBN 978-1-804515-68-6

British Library Cataloguing-in-Publication Data.
A catalogue record for this book is available from the British Library.

## Contents

## List of Illustrations

## List of Maps

## List of Abbreviations

| | |
|---|---|
| ADC | Aide de Camp |
| ADMO | Assistant Director of Military Operations |
| APS | Army Postal Service |
| ASC | Army Service Corps |
| BEF | British Expeditionary Force |
| BGGS | Brigadier-General, General Staff |
| CID | Committee of Imperial Defence |
| CIGS | Chief of the Imperial General Staff |
| DAAG | Deputy-Assistant Adjutant General |
| DMO | Director of Military Operations |
| DMT | Director of Military Training |
| GHQ | General Headquarters |
| GOC | General Officer Commanding |
| GSO | General Staff Officer |
| IGS | Imperial General Staff |
| LoC | Line of Communications |
| MI | Mounted Infantry |
| MO | Medical Officer |
| NCO | Non Commissioned Officer |
| PSC | Passed Staff College |
| QMG | Quarter-Master General |
| RAMC | Royal Army Medical Corps |
| RFA | Royal Field Artillery |
| RFC | Royal Flying Corps |
| RGA | Royal Garrison Artillery |
| RHA | Royal Horse Artillery |
| RUSI | Royal United Services Institution |
| WO | War Office |

# Foreword

The 21st century has been a remarkable time to be a military historian of the First World War. Recent years have witnessed an outpouring of fascinating research which has shed new light upon innumerable aspects of the conflict. A common theme of much of the work studying the British Army has been a desire to explore the concept of the 'learning curve' as originally proposed by Professor Peter Simkins. As noted in the introduction to this volume, the study of how the army learned from its experiences has demolished the disillusioned caricatures of lions led by donkeys. It is now widely accepted that – severe casualties notwithstanding – the British Expeditionary Force showed an impressive capacity to adapt, innovate and learn at tactical and technical levels. By the end of the war a case can be made that the BEF was the most advanced and effective army in the field.

An interesting undercurrent to the 'learning curve' debate is the fact that Field Marshal Sir Douglas Haig attributed victory to following the principles of the pre-war *Field Service Regulations* manuals. Haig's belief in the value of his pre-war military education has been noted by a range of biographers and has been used as both a compliment and a criticism. Whilst a growing body of modern scholarship has found much to praise in the pre-war army's flexible approach to battle, the extent to which the British Army was prepared for high intensity warfare in 1914 remains contentious.

Much of this debate centres around the training undertaken by the pre-war British Army. Training is an essential component of military success. An oft-repeated quote, usually attributed to the Japanese swordsman Miyamoto Musashi, sums the matter up: 'We do not rise to the level of our expectations, we fall to our level of our training.' Historians have praised the pre-war army's small unit tactics and noted its exceptionally high standard of marksmanship. Yet there is a sense that the Army relied too much on its battalions to solve battlefield problems and simply fight their way out of trouble; an attitude embodied by the motto of the day "We'll do it. What is it?". At higher levels of command there were fewer opportunities for development and it has been suggested that the senior officer class was ill-prepared for high intensity warfare.

Simon Batten's work is therefore immensely valuable as it offers the first serious study of how the British Army trained for large scale operations. One may perceive here the very beginnings of the 'learning curve' as the army began the painful reforms that were required after the Anglo-Boer War. Manoeuvres were still a novelty and the army had to learn how to organise and umpire them efficiently. Efforts in this regard were hampered by inexperience and limited by time, space and budget. Yet the overall improvement in the conduct of manoeuvres was marked. In 1898 the Army's manoeuvres had been farcical but by the eve of the First World War they had developed into a recognisably modern exercise which incorporated a preliminary staff ride, large scale troop deployment, and post-action reflection. At their best these manoeuvres could provide valuable lessons on timings, logistics and the employment of service arms, particularly the fledgling aviation branch.

Nevertheless, Batten makes clear that manoeuvres were not a universal panacea to the problems of command. Exercises could be impractical, and realism was perennially difficult to achieve. How much was learned on manoeuvre is difficult to quantify. Some officers, including Douglas Haig and Edmund Allenby, found themselves repeating the same errors they had made in peacetime exercises during the opening weeks of war. Others who had done well in manoeuvres, notably Ivor Maxse, found that their expertise availed them little during the desperate days of August 1914.

Yet whilst some individual officers struggled, the fact that the BEF survived the exceptionally harrowing Great Retreat and was then able to turn and fight at the Battle of Marne surely owed something to the experience gained during fighting withdrawal exercises in 1913 and 1914. Indeed, as Batten ably demonstrates, manoeuvres were a positive influence upon the development of the British Army. They gave experience in solving many practical problems, prompted reflection on the nature of operations and contributed to a culture of professionalism. As the author rightly concludes, these were far from a 'futile exercise'.

Dr Spencer Jones
*University of Wolverhampton*

## Acknowledgements

My greatest debt is to Spencer Jones, whose work on the British Army's tactical development following the Boer War helped to inspire this book and who generously agreed to write the foreword as well as continually offering advice and encouragement throughout the long process of research and writing. My father, Lt. Col. Frank Batten, was an important source of ideas, even if he has not always agreed with some of my conclusions. I am grateful to John Hussey for acting as a wise mentor who has been generous with his time and invaluable in pointing me in the direction of new avenues of research, while Walter Reid was similarly helpful, especially when it came to his fellow Scots, Generals Haig and Grierson.

A first time author, and especially one writing in isolation away from a university department, must incur a great many debts in the course of his research, and I have been overwhelmed by the generosity and readiness to offer advice exhibited by a large number of historians. Prominent among these have been John Bourne, Gary Sheffield, Andrew Lambert, Holger Afflerbach, Bill Philpott, John Lewis-Stempel, Richard van Emden, Jim Smithson, Taff Gillingham, Simon House, John Lee, Peter Rostron, Bill Thompson, and David Walsh. I am extremely grateful for Nick Evans' constructive criticism of some of the points made in the article which was a precursor to this book, allowing me to refine some of my initial ideas. Similarly, Andrew Winrow pointed out the performance of the Mounted Infantry in 1912, to which I had paid insufficient attention.

David Morgan-Owen was the source of a great deal of good advice, especially when it came to amphibious warfare and invasion literature, and was one of several historians who reassured me that the question I had posed was a valid one. Margaret Webb assisted me greatly on the 1913 manoeuvres, and José Madroñero was extremely helpful with Canadian archive material. I have also derived regular inspiration from Stephen Badsey, Phylomena Badsey and the First World War study group at the University of Wolverhampton. I learnt a great deal from battlefield tours to Mons and Le Cateau with Mike Peters and Clive Harris, to Ypres and the Somme with Mike Sheil and David Winn and to Ypres with Matthew Dixon. My former tutors, John Walsh and Felicity Heal, offered invaluable encouragement and advice as ever. I will always be grateful for their inspirational teaching, along with that of Edward Parry and Glyn Edwards.

My thanks go to the staff of the following libraries and archives: the Bodleian Library, the National Archives, the Imperial War Museum, the Liddell Hart Centre for Military Archives, the National Army Museum, the Coldstream Guards Archive, Cambridge Central Library, Clacton Library, West Clacton Library, and the National Library of Scotland. I would like specifically to thank Major Robert de L. Cazenove (Archivist, Coldstream Guards), Alison Metcalfe of the National Library of Scotland, David Martin of the Little Shelford History Society, Andrew Westwood-Bate of Hildersham History Recorders, Tony Turner and the Haverhill & District Local History Group, Roger Colbourne of Bugbrooke Link, Rupert Clark and the Aynho History Society, Anthony Benn of the Chipping Norton U3A, Clare Kavanagh (Assistant Librarian of Nuffield College, Oxford), and Jonathan Smith (Archivist, Trinity College, Cambridge). I am also grateful for advice on specific points to Andrew Cormack, Tim Moore (on contacts between the French and British staffs), Derek Linney (invasion literature), Robert Batten (Evelyn Wood), Marc Batten (physical training in the British army), William Fletcher, Toby Clark, and Alex Young.

I would like to express my thanks to the Headmaster and Governors of Bloxham School for granting me a sabbatical term in 2013, when the seeds of this work were sown. I am grateful to my colleagues for help and encouragement, and especially to Dr Debbie Herring and my colleagues in the History Department, Ian Hatton and Robert Hudson. Danielle Sterrenburg provided assistance in Dutch translation and Andrew Whiffin did a fine job of proof-reading. It goes without saying that any errors which remain are my responsibility alone.

I am grateful to all those who helped with the sourcing of images, in particular Sonia Nicholson of the Saanich Archives in British Columbia, who drew my attention to the wonderful photographs of the 1912 manoeuvres in Owen Aspray's collection. I am indebted to Tim Moore who drew my attention to the two splendid Punch cartoons and Andre Gailani of Punch Limited who enabled me to use them for this book. George Anderson has provided some excellent maps. Finally I would like to express my appreciation for the faith shown in this project from the very start by Duncan Rogers and Helion. My sincere thanks go to Duncan and to Michael LoCicero for their support and advice, and above all their forbearance in dealing with a novice in this field.

## Introduction

---

According to the official historian, Brigadier-General James Edmonds, 'In every respect the Expeditionary Force of 1914 was incomparably the best trained, best organised, and best equipped British Army which ever went to war.'[1] This assertion (which was, crucially, followed by the author going on to contradict himself by listing a variety of areas in which the BEF was sadly deficient) has been the starting point for many historians when examining the British performance at the outbreak of the First World War.[2] There has been considerable debate over the extent to which Edmonds's claim was justified and whether the British Army had learnt the lessons of recent events, above all its chastening experiences in South Africa.[3]

As a teacher with over thirty years of experience of coaching schoolboy rugby, I have long been preoccupied with the question of how one should best prepare a sports team for a match. Typically, a rugby side has an unopposed practice on a Friday, with the players running through their moves without opponents to get in the way and invariably looking slick and assured in the process, only for things to go awry on match day when confronted with opponents who tackle them and with situations which develop unpredictably. Under the pressure of competition, mistakes are made and things which worked well in the practice the day before go badly wrong. This observation and my interest in military history made me question how armies practise for war, and whether the circumstances of exercises which inevitably do not entail the use of live ammunition can possibly provide any worthwhile preparation for real combat. The comparison between sport and warfare is a familiar, not to say hackneyed one, especially when it comes to the First World War,[4] and obviously there are huge differences between being tackled in rugby and being shot at in combat, but the point about the gap between practice and reality, and the role of the Clausewitzian concept of 'friction' is surely a valid one. Despite what Edmonds said about the army of which he was a part in 1914, there has been widespread agreement that the British Army

1 Brig. Gen. J. Edmonds, *Military Operations France and Belgium 1914* pp.10-11.
2 For example, John Bourne, 'British Readiness for War' in Peter Liddle, *Britain Goes to War: How the First World War Began to Reshape the Nation, p.23.*
3 S. Jones, *From Boer War to World War: Tactical Reform of the British Army, 1902-1914.*
4 See P. Parker, *The Old Lie, The Great War and the Public-School Ethos*, pp.211-217, A. Seldon and D. Walsh, *Public Schools and the Great War*, pp.21-22.

was not ready for the conditions it faced in the First World War. This was despite the previous decade witnessing a growing consensus among the high command that war was on the way, that Britain would be involved and that Germany was Britain's likely enemy. The British army had carried out repeated exercises since the end of the Boer War designed to prepare for war and yet, when that war came, the army was apparently not ready.

In order to assess the extent to which manoeuvres helped to prepare the British Expeditionary Force for the test it would face in 1914, this book will examine the British Army's preparations for war between the end of the Boer War and the outbreak of war in August 1914 and will go on to look at the first months of the war, when the BEF found itself fighting alongside the French and Belgians in attempting to resist the German advance on Paris. The development of trench warfare from October 1914 onwards will be taken as the terminal point, as from that moment the war ceased to resemble the war anyone had prepared for. Even so, events from later in the war will be alluded to where there is a connection with pre-war developments. The main focus will be on the ten sets of army manoeuvres held in September of each year between 1903 and 1913, although reference will also be made where relevant to other elements of the Army's 'Collective Training Period', including divisional and brigade manoeuvres, staff rides and tactical exercises. In order to avoid tedious repetition, three sets of manoeuvres will be studied in detail, those of most interest and relevance, each of which took place away from their usual location on and around Salisbury Plain, namely those of 1904 (Essex), 1912 (East Anglia) and 1913 (Buckinghamshire and Northamptonshire).

Army manoeuvres could be massive affairs, involving up to 50,000 troops and many of the generals whose names would soon become familiar from their actions on the Western Front, such as Haig, French, Gough, Rawlinson and Allenby. These men have generally suffered a bad press, portrayed as 'donkeys' and presented as obstinate and unimaginative old men, sending hundreds of thousands of brave young men 'over the top' to their deaths, hurling them against machine guns in a series of futile attacks while they themselves were skulking safe in their chateaux, miles behind the lines.[5] Every element of this caricature has been convincingly dealt with, at least to the satisfaction of academic historians, but among the general public it still shows every sign of being what the Australian historian Craig Stockings has termed in a different context a 'zombie myth' which refuses to die.[6]

British generals of the First World War were not particularly old – Sir Douglas Haig was 54 when he took over command of the BEF at the end of 1915. Many of them were imaginative and ready to embrace new technology whenever it offered some

5 For an extreme version of this view, see J. Laffin, *British Butchers and Bunglers of World War 1.*

6 Notable examples of this revisionism are J. Terraine, *The Smoke and the Fire: Myths and Anti-Myths of War 1861-1945* and G. Sheffield, *Forgotten Victory. The First World War: Myths and Realities.*

prospect of an advantage over the enemy. Seventy-eight of them died in action, which would appear to disprove the 'chateaux generals' myth, the youngest of them being Roland Boys Bradford VC, killed at the age of 25 at Passchendaele in 1917. It cannot be denied that the British losses on the Western Front were enormous, as were those of other combatant nations, but with the German army occupying the soil of Britain's French and Belgian allies, the British Army had to attack, even if attack was very difficult in the new conditions of warfare imposed by trenches, barbed wire, machine guns and, above all, rapid-firing heavy artillery. As the French General Mangin observed, 'whatever you do, you lose a lot of men.' The British Army had actually become by 1918 a massive and extremely effective fighting machine which had learnt from the lessons of 1915 and 1916 and was able to use an integrated weapons system, employing tanks, short bombardments, mobile infantry and aerial support which delivered arguably the greatest series of victories in British history in the last hundred days of the war. The British Army, along with its allies, won and the much-vaunted Germans (whose generals were on average four or five years older than the British) lost.

But after all this revisionism a major criticism remains, that British generals were slow to adapt to the new warfare which they encountered in 1914, even though they had encountered rapid-firing heavy artillery in the Boer War (1899-1902), they knew of the destructiveness of machine guns in the Russo-Japanese War (1904-1905) and many of them had studied at Staff College the American Civil War (1861-1865), which had featured trenches and attritional positional warfare in its final stages. Why weren't they better prepared? How had they prepared? How can an army prepare for something as unknowable and incapable of simulation as warfare?

Until the present author's article on the 1912 manoeuvres, D.M. Leeson's study of those held in 1898 was the only modern operational study of a set of army manoeuvres of the period, and that falls outside the chronological scope of this book.[7] In addition, John Sunderland and Margaret Webb have studied the 1913 exercises from the point of view of their impact on the area in which they took place, although they also relate these events to those of 1914. Otherwise, the response of most historians to army manoeuvres has been to ignore them or use them as a dependable fund of humorous anecdotes to illustrate the supposed amateurishness of the pre-war British Army.

Were these manoeuvres worthwhile? The responses to my question from two men who commanded infantry battalions in the Cold War era are typical:

> I have never been of the opinion that you can really test commanders and staff by deploying vast numbers of men and *materiel* around the country, nor is it economical in time and money. Thankfully, someone devised TEWTs (Tactical Exercises without Troops).[8]

7 S. Batten, 'A school for the leaders: What did the British Army learn from the 1912 army manoeuvres?' *Journal of the Society for Historical Research,* pp.25-47.

8 Lieutenant Colonel F.C. Batten, Royal Regiment of Wales.

> Large-scale exercises are certainly desirable to give senior commanders and staffs practice at handling men and *materiel.* But I do not believe that they give a reliable judgement on commanders' tactical abilities.[9]

A number of serving army officers with whom I have discussed the subject have also questioned why the British bothered to hold laborious, expensive and time-consuming manoeuvres in the first place, given their inherent lack of realism. I hope that I may be able to change their minds in this book. In order to avoid wearisome repetition, I have avoided putting quotation marks around words such as 'battle', 'losses', and 'victory' each time, even though I recognise the difference between real warfare and mock warfare (although it should be recorded that these manoeuvres were not entirely bloodless).

9 Colonel P. Rostron, Gloucestershire Regiment.

# 1

# The British Army and Manoeuvres

Manoeuvres, which were defined in the 1911 Encyclopaedia Britannica as 'the higher training for war of troops of all arms in large bodies' had been in use since the formation of standing armies, and Louis XIV had employed large camps of instruction, though the manoeuvres performed there were of a basic manner and much of the time was devoted to ceremonial.[1] The first army to use manoeuvres in a systematic fashion was that of Prussia; 36,000 men carried out manoeuvres at Spandau in 1753 for twelve days. These were held under conditions of great secrecy, with no strangers allowed to be present. Frederick the Great used such exercises to develop on the training-ground the complex system of tactics which proved so successful in his campaigns. The results of all this training were to be seen in Frederick's victories in the Seven Years' War, and this led to Prussian manoeuvres gaining a reputation which would endure for more than a century, though the Prussian army's failure to move with the times after Frederick's death resulted in an ossified military system which was rapidly dealt with by Napoleon in 1806. After the Napoleonic Wars, yearly manoeuvres became the custom in every large continental army. Carl von Clausewitz (1780-1831) was clear about the value – and limitations – of peacetime manoeuvres:

> No general can accustom an army to war. Peacetime manoeuvres are a feeble substitute for the real thing but even they can give an army an advantage over others whose training is confined to drill. To plan manoeuvres so that some of the elements of friction are involved, which will train officers' judgement... is far more worthwhile than inexperienced people may think.[2]

The British lagged some way behind continental armies in making use of manoeuvres and large-scale exercises, with the first attempts not being made until the start of the nineteenth century, when Sir John Moore trained the Light Brigade at Shorncliffe

1 Lt. Gen. Sir J. Grierson, 'Military Manoeuvres', *Encyclopaedia Britannica,* p.592.
2 C. von Clausewitz, eds M. Howard & P. Paret, *On War,* p.122.

camp. It was not until 1853 that a body of troops was gathered together for a camp of exercise on Chobham Common, and it took several more years before the temporary camp formed at Aldershot during the Crimean War became a permanent one. The full-scale manoeuvres held at Aldershot in 1871, staged against a backdrop of disquiet at Prussia's stunning defeat of France, were on an unprecedented scale, and led one young officer to assert that they marked Britain's coming of age as a military power.[3] Further manoeuvres were held in 1872, 1873 and 1875, but after that there were no more large-scale exercises until the 1890s, ostensibly on the grounds of cost.[4]

The Prussian victory in 1871 meant that other continental armies copied German training methods wholesale, but it was not until the end of the century that Britain made any serious attempt to catch up with the Germans. Even after this, there was a recognition that Britain still lagged behind Germany in terms of the scope for independent action given to subordinate commanders in manoeuvres. Major Frederick Trench observed in 1894, 'To anyone fresh from our own camps of exercise there is perhaps nothing more striking at German manoeuvres than the amount of independence and initiative allowed to the commanders of the smaller units.'[5]

Two developments were central to the process by which the British properly embraced the use of manoeuvres, the passing of the Military Manoeuvres Act of 1897 and the purchase of land on Salisbury Plain to provide a training ground for the army. The Military Manoeuvres Act gave the government the power to 'proclaim' the use of any district it wished for manoeuvres, provided an order had been published in local newspapers not less than three months before the order was to come into force. Roads could be closed, subject to due notice being given and consent from Justices of the Peace, and the Act provided for penalties for wilful and unlawful obstruction of manoeuvres. This was a significant step forward, although there were severe restrictions on the army's access to farmland and woods, and the fact that the Act did not give the army the power to billet troops on private homes, as the Germans could, was a source of frustration to many soldiers and politicians who felt that manoeuvres were still not being taken sufficiently seriously. The Army Act was amended in 1909 to allow troops to be billeted in an emergency but this provision was little used and the Army preferred to trial a voluntary scheme, first used during the cavalry divisional training in Dorset and Wiltshire in August 1910, whereby farmers were invited to provide overnight accommodation, usually in barns, for a payment of one shilling for each man and horse.[6]

By the early 1890s the British Army was finding the area around Aldershot to be too restricted for its purposes. Housing development in the area was preventing further

3 'The Autumn Manoeuvres of England', *Journal of the Royal United Services Institution* (1873), p.236.
4 Howard Bailes, 'Technology and Tactics in the British Army, 1866-1900' in R. Haycock, K. Neilson, *Men, Machines and War*, p.37.
5 H. Bailes, op. cit., p.28.
6 T. Crawford, *Wiltshire and the Great War: Training the Empire's Soldiers*, p.19.

Army manoeuvres, 1894 – volunteers attacking regulars at Aldershot. (Mary Evans Picture Library)

expansion and driving up house prices. A more extensive training area was required for infantry and horse artillery, and cavalry needed an area of flat, open grassland. There was also a pressing need for ranges long enough to allow practice with the new Lee-Enfield rifle. Thanks to the Military Lands Act of 1892, the Government now had the power to acquire land for military purposes, and in January 1897 the decision was taken to spend £450,000 on buying 40,000 acres of Salisbury Plain for manoeuvres. The first land near West Lavington was purchased in August, and by the end of the year more than 13,600 acres had been bought. By March 1900, 42,000 acres of the eastern and central areas of the Plain had been purchased, costing the Government £560,000.[7] The eastern section was used mainly for manoeuvres, cavalry exercises and rifle ranges, while the area north of Amesbury was allocated for artillery practice.

The first army manoeuvres on the new land took place in September 1898, when two Army Corps, each comprising three infantry divisions and a cavalry brigade, a combined total of 50,301 men, 9450 horses and 226 guns, fought a week-long campaign which ended in a humiliating defeat for Sir Redvers Buller's Southern army (Blue) at the hands of the Red army of the Duke of Connaught. Buller's performance was particularly embarrassing given his status as the protégé of the Commander in Chief, and the rising star of the late Victorian army. His repeated frontal assaults on

7 Ibid., p.12

the Red forces were deemed by the umpires to have been costly failures. Men of Sir William Butler's 4th Division stumbled forward in close formations into enfilading machine gun fire. The correspondent of *The Daily Chronicle,* who was 'embedded' with Buller's army, reported that 'it was a helpless, hopeless massacre of a whole brigade. Nothing could have saved us.'[8]

Artillery fought in the open, exposing itself to counter-battery fire, and Sir Arthur Conan Doyle wrote that 'In the Salisbury Plain Manoeuvres of 1898 I saw with my own eyes lines of infantry standing and firing upon each other at short ranges without rebuke either from their officers or from the umpires.'[9] D.M. Leeson, in his damning study of these manoeuvres, has argued that they reveal the late Victorian army to be utterly undeserving of attempts by revisionist historians to portray it as a reforming force developing a modern school of tactics and adjusting to new conditions of war: 'the manoeuvres of 1898 show little evidence of either tactical ideas or training.'[10] Leeson goes on to argue that given the events on Salisbury Plain in September 1898, the poor performance of the British Army in the Boer War should come as no surprise: 'When examined in light of the manoeuvres of 1898, the disasters of 1899 appear almost inevitable.'[11]

By choosing to base his argument on a notoriously badly-handled set of manoeuvres Leeson was certainly able to construct a persuasive case, and it is not hard to find evidence of almost laughable ineptitude in the manoeuvres of this period. At the 1896 Aldershot manoeuvres, one battery opened fire on a hill packed with spectators, while in 1895 a battalion from the Northern Force managed to charge its own artillery. What this book sets out to examine is whether there was any improvement in this state of affairs following the Boer War in the lead-up to the First World War, and this can be done through a study of the ten sets of army manoeuvres which took place between 1903 and 1913.

There was widespread acknowledgement that things had to change in the years following the South African War. This sentiment was articulated most clearly by Major-General James Grierson, who wrote from Pretoria on 7 July 1900 to Arthur Bigge, Queen Victoria's Private Secretary, in the following terms:

> What a lot we have to learn from this war in every way! I think our first lesson is that we must have big annual manoeuvres and have our staffs properly trained. We don't seem to grasp anything higher than a division. And we must have 'staff

8 *The Daily Chronicle,* 3 September 1898.
9 Sir A. Conan Doyle, T*he Great Boer War*, p.517.
10 For examples of this revisionism, see H. Bailes, Technology and Tactics in the British Army, 1866-1900 in R. Haycock, K. Neilson, *Men, Machines and War*; G. Phillips, 'The Obsolescence of the *Arme Blanche* and Technological Determinism in British Military History', *War in History.*
11 D.M. Leeson, 'Playing at War: The British Military Manoeuvres of 1898', *War in History,* p.434.

> journeys' to teach the control of armies in the field. If we take the field with a force the size of this one against a European enemy and continue in our present happy-go-lucky style of staffing and staff work we shall come to most awful grief.[12]

The majority of these army manoeuvres were held in a small area of the West Country, centred on Wiltshire and the new training ground on Salisbury Plain. The 1905, 1906, 1907 and 1910 manoeuvres all took place there, and so would those of 1911 have done if they had not been cancelled due to drought. In 1903, the first set of manoeuvres after the end of the Boer War, the area involved extended beyond Salisbury Plain, stretching as far as Hampshire, Oxfordshire and Berkshire, and again in 1908 (centred on Hampshire) and 1909 (between Cheltenham, Swindon and Oxford). One French observer, General Langlois, noted in 1909 that the plain was more undulating and contained a greater number of clumps of trees than the French manoeuvre grounds at Châlons-en-Champagne.[13] However, as early as 1904, concerns were being expressed that the Army needed to practise in terrain that was more typical of the more wooded and less flat ground that they might encounter in resisting an invasion of south-east England. In justifying the choice of East Essex for the 1904 manoeuvres, the Chief of the General Staff, General Sir Neville Lyttelton, pointed out that 'Hitherto, manoeuvres have taken place at Aldershot, on Salisbury Plain, and on the Berkshire Downs, where the country is open and where very marked positions exist… Our Army had, hitherto, very rarely or never been exercised in operations in a close country such as the County of Essex presents.'[14] This concern was even more evident in the last few years before the outbreak of war, as attention shifted from the danger of invasion to the likelihood of an expeditionary force being committed to the European mainland. Most commanders were better informed than Colonel Beauvoir de Lisle, who exhibited some confusion when informing the audience for his lecture at Aldershot in February 1911 that:

> I have not been to Belgium myself, but everyone has assured me the ground there is very like Salisbury Plain, with a few farms here and there, no hedges and a few ditches… Such ground is more suitable for cavalry than you could find anywhere else in Europe.[15]

Within four years, de Lisle would have ample opportunity to see the error of his judgement. The 1912 manoeuvres (Cambridgeshire and Suffolk) and those in 1913

12 D. Macdiarmid, *The Life of Lieut. General Sir James Moncrieff Grierson*, p.271.
13 General H. Langlois (translated by Captain C.F. Atkinson*), The British Army in a European War*, p.17.
14 The National Archives (TNA) WO 279/8, Report on Army Manoeuvres 1904, p.16.
15 Lord Anglesey, *A History of British Cavalry 1816-1919 Volume VII: The Curragh Incident and the Western Front 1914*, p.78.

(Northamptonshire and Buckinghamshire) both took place in countryside more closely resembling that of Belgium and northern France than Salisbury Plain did.

For many years the British Army had been perceived as failing to take manoeuvres seriously, not least in terms of their size. British Army manoeuvres were not negligible affairs – even back in 1872, there had been 30,000 troops engaged, and those of 1898 had involved 50,000 soldiers, a far larger force than was engaged in the same week at the battle of Omdurman.[16] However, British army manoeuvres were still on a smaller scale than those of its European rivals. The 1912 manoeuvres involved only 51 infantry battalions and 33 cavalry squadrons whereas those in Germany the previous week, held in Saxony, involved 110 battalions and 129 squadrons (the 1912 Russian equivalent featured 112 battalions).[17] Colonel Cecil Lowther, a regular attender of continental manoeuvres, expressed the fear that 'it would be idle to pretend that we even begin to understand the handling or movement of large bodies, for the very sufficient reason that we have no large bodies to handle.' Lowther observed in 1911 that 'We now make quite a fuss if we turn out two whole divisions on manoeuvres; the French and other great Continental nations have annual manoeuvres including two or more army corps, and do not think they are doing anything very big.'[18]

Having said that, such was the small size of the pre-war British army that the 50,000 troops involved in the 1910 manoeuvres represented almost half of the entire forces available. In comparison, the manoeuvres staged by the mass conscript armies of France and Germany involved a much smaller proportion of their land forces. Given the prevalent tendency to assume German military superiority, it is worth noting that the Chief of the German General Staff, Field-Marshal Helmuth von Moltke, expressed the fear that engaging two or three corps in army manoeuvres every autumn did not provide him or anyone else in the German Army with actual experience of commanding millions of men and dozens of corps as they would when war came – in 1914 the three armies in the German right wing alone totalled 700,000 men in sixteen corps.[19]

Therefore, British commanders in their manoeuvres were actually commanding forces much closer in size to those they would lead when war came than their continental counterparts; Field-Marshal Sir John French led an army of 38,000 men in the 1913 manoeuvres compared to an expeditionary force of 80,000 in August 1914, while Grierson's Blue Army of 25,000 men in 1912 was not much smaller than the II Corps (36,000 men) he took to France when war broke out. An understandable tendency on the part of British historians to give a prominence to the BEF's actions at Mons and Le Cateau out of all proportion to their significance in the bigger picture

16 Leeson, op. cit., p.433.
17 TNA WO 33/618, Report on Foreign Manoeuvres 1912.
18 Colonel H. C. Lowther, *From Pillar to Post*, p.167.
19 E. Dorn Brose, *The Kaiser's Army: The Politics of Military Technology in Germany during the Machine Age, 1870-1918*, p.158

'How to War without causing inconvenience', *The Graphic*, 21 September 1912. (Author's collection)

should not blind us to the fact that the British military presence on the continent in August 1914 was miniscule in comparison with the armies of France and Germany. The Germans put 1,077 battalions into the field in the west in 1914, the French 1,107 and the Belgians 120. The British initially fielded just 48.

Even after the 1897 Act and the acquisition of Salisbury Plain, practical problems limited the usefulness of the manoeuvres process. The difficulty of finding suitable land is illustrated by the experiences of Eastern Command, which could not call upon Aldershot or Salisbury Plain and was perennially short of land on which to train. It therefore sought to hire an area of about 40,000 acres near Lewes for brigade and divisional training between June and August 1906. The Army Council received an application to have a tract of land of about 40,000 acres placed under the provisions of the Manoeuvres Act but, faced with concerted opposition from local landowners, sporting tenants and farmers, Eastern Command was forced to scale down its plans and limit the training to one month, August. Aldershot Command encountered similar problems when Smith-Dorrien decided that operations should include a night crossing of the Thames by a division. A location for an improvised bridge was selected, and the ground had been obtained on one side of the river, only for the owner of the fields on the opposite bank to refuse point blank to co-operate. It took a lengthy correspondence and a personal visit from William Robertson before the landowner's objections were overcome.[20]

Training also suffered from a consistent shortage of troops. In 1913 Sir John French in theory commanded four infantry divisions and a cavalry division, a force which would have totalled 81,500 men in terms of its 1914 establishment, but in practice only 47,000 men took part.[21] Battalions were, on average, less than half their war-time establishment. The staggering of the training year, which made sense in other respects, had unfortunate consequences. Because the British volunteer system, unlike the German and French, took in new men throughout the year, men often arrived at their companies with little training. Brigadier-General Maxse declared in an address to the RUSI in 1911 that most company commanders 'begin to realise that the new comers are raw recruits totally ignorant of field work, and with a large proportion of these [they proceed] to army manoeuvres.' He asked his audience, 'What company officer in a foreign army, where all recruits join the same day, has such a difficult task?'[22] According to Maxse, company strengths averaged about 40 men because of officers and men being detached for a whole variety of other duties, such as subalterns being off on courses during the training season.

There was the added problem of having to find drafts for overseas units, especially India. Major-General S.C.M. Archibald of the Royal Artillery complained that 'Most of one's best NCOs and men seemed to be taken away from one every year and replaced by a lot of untrained recruits from whom one was expected to build the battery again by the spring into a first-class show.'[23]

There was a particular difficulty with members of the Territorial Force, introduced by Lord Haldane in 1908. Within a year of its creation the Territorial Force had reached 9,313 officers and 259,463 other ranks, but recruiting declined after that, and concern was soon being expressed that numbers were inadequate for its stated role of home defence.[24] At the 1908 manoeuvres, one brigade had allocated posts for 3,840 men, excluding officers, but had only 1,287 on strength, of whom 1,043 were in camp on Salisbury Plain. Even then, 45% only attended for eight days rather than the full fortnight – part of the problem was that employers were unwilling to release workers for longer. Some employers were extremely supportive of the new scheme, offering to raise units from their employees and to provide training facilities on their premises. Local studies have shown how the picture varied markedly across the country, as well

20 Field-Marshal Sir W. Robertson, *From Private to Field-Marshal*, p.162.

21 E. Spiers, 'Reforming the Infantry of the Line 1900-1914', *Journal of the Society for Army Historical Research*, p.87.

22 T. Bowman and M. Connelly, *The Edwardian Army: Recruiting, Training, and Deploying the British Army, 1902-1914* p.67.

23 Ibid.

24 P. Dennis, *The Territorial Army 1907-1940*, p.17.

as between large and small employers.[25] Colman's of Norwich offered support for the Territorials, but only for one hundred of their employees. While W.D. & H.O. Wills of Bristol promised Territorials a full twenty-one days' annual leave on full pay in order to attend camp and training, and Birmingham employers – the BSA, Dunlop Rubber Company and Mitchell & Butler – offered full support, elsewhere employers were less co-operative.[26]

By 1914 the situation had improved, thanks to a bonus system designed to encourage Territorials to stay for the full fortnight, but many were still afraid that they would lose their civilian jobs if they did not get back to work on time.[27] It did not help that the Trades Union Congress was unconvinced about the Territorial scheme, and there was also concern over the absence of any insurance scheme for men serving with the Territorials – this issue was brought into sharp focus by the plight of the widow and children of Gunner Stone of the London Territorials, who died at annual camp at Bulford in August 1909.[28]

Other practical problems were encountered: it was impossible to get all six divisions of the army together as 6th Division was needed for policing duties in Ireland. Financial considerations, which precluded the use of Scotland and Ireland since all but two of the army's divisions were in England, also severely limited the availability of ammunition. The annual allowance for training purposes, granted in 1902-3, had been raised from 150 to 300 rounds per man for cavalry and from 200 to 300 for infantry, but this ended when stocks of wartime ammunition ran out. When these costs threatened to increase from £70,000 to £80,000 the new Liberal Government, committed to keeping the Army Estimates down, reduced the allowance to 250 rounds per year despite protests from its military advisers.

This was still a far higher allowance than a German infantryman, who was given between 60 and 100 rounds in his first year and 42 in his second year, but British commanders argued that the reduction had serious implications for troops on manoeuvres, with concern being expressed in 1909 that there was insufficient blank ammunition to test out the effects of greatly increased firepower on the battlefield.[29] Brigadier-General Rawlinson argued that 'it is very difficult to train troops to take a really keen interest in establishing superiority of fire, unless they have sufficient blank ammunition to keep up the fire.' Rawlinson complained that 'by the time they get launched in an attack they soon fire off their small supply of ammunition, and superiority of fire degenerates into "snapping", which gives no visible result, and fails to make any impact on the umpire.'[30] At the 1910 manoeuvres the allocation of blank

25 A. Thornton, *The Territorial Force in Staffordshire, 1908-1915*, p.29.
26 I. Beckett, *Territorials, A Century of Service*, p.34.
27 Crawford, op. cit., pp.18-22.
28 Thornton, op. cit., p.31.
29 Spiers, op. cit., pp. 87-88.
30 TNA WO 279/25, Report of a Conference of General Staff Officers at the Royal Military College 1909, p.10.

ammunition was 50 rounds per man for infantry and 25 for cavalry, 240 rounds of smokeless powder per battery for the RHA and RFA, 120 rounds of black powder per battery of RFA howitzers and 120 rounds per battery of RGA heavy guns. Machine guns received 200 rounds each.[31]

Manoeuvres could be cancelled due to a variety of outside events, such as an outbreak of foot and mouth (1912 in Ireland) or a strike (the 1911 divisional manoeuvres). Even the coronation of King George V could disrupt training through the need for large numbers of soldiers on ceremonial duties. Most significantly, with the divisional manoeuvres having already been impeded by the national railway strike, as troops were engaged in guarding railway installations and coping with riots, the 1911 army manoeuvres were cancelled due to 'scarcity of water in Wiltshire'. That, at least, was the ostensible reason; *The Salisbury Times* was not alone in suggesting that the real reason for the cancellation was the unsettled international situation resulting from the Morocco Crisis.[32]

Landowners were reluctant to rent out their land for training purposes and invariably submitted inflated demands for compensation after the manoeuvres were over. Troops had little incentive to dig trenches, knowing that the deeper they dug, the more earth would have to be filled in again afterwards. In order to avoid heavy claims for compensation, entrenching was not permitted in areas with a chalk subsoil close to the surface, a not uncommon situation in southern England – as would also later be the case on the Somme battlefield. Where the ground was not to be broken, troops were supposed to carry tape along with their entrenching tool so that the site of the trench could be marked out.

One umpire at the 1909 manoeuvres – unsurprisingly, a Royal Engineer – noted that 'much has to be done to make troops believe that good digging powers are as important as good marching powers', to which the Blue Umpire-in-Chief Douglas added laconically that 'this is probably overstated.' Douglas did, though, concede that 'there appeared to be a marked disinclination on the part of the troops to dig trenches.' Soldiers preferred the 'exhilarating' business of advance and attack to 'the dull work of digging'.[33] Lieutenant-General Douglas Haig was shocked by what he saw on an inspection of his Aldershot command in March 1912: 'Ride out at 9:00 a.m. See a company of Bedfords entrenching. Several with coats on! Many could not use pick and shovel.'[34]

There was, in any case, little official encouragement to dig deep trenches. The *Manual of Field Engineering* (1911, reprinted in 1914) made it clear that trenches were to be seen always as a 'means to an end', designed to enable troops to use their weapons

31 TNA WO 279/31, Special Instructions contained in Report on Army Manoeuvres, 1909.
32 Crawford, op. cit., p.21.
33 TNA WO 279/31, Report on Army Manoeuvres 1909, p.112.
34 Douglas Haig (ed. Douglas Scott), *The Preparatory Prologue: Diaries and Letters 1861–1914*, p.143.

Trench digging at Horseheath, September 1912. (Saanich Archives)

to the 'greatest effect'. There was a recurrent fear that if attacking troops dug in too enthusiastically, they would be reluctant to leave the security of the trenches and the impetus of the attack would be lost. Major-General Altham warned in his *Principles of War* (1914) that any force having perpetual recourse to the spade 'would destroy its morale, encourage the spirit of the enemy, and wreck whatever plan of offence may have been formed by supreme authority.'[35] As in the German instruction manual *Feld-Pionierduenst Aller Waffen* (Field Pioneer Work for all Arms, 1911), British manuals were keen to stress the primary role of trenches as providing a fire position and a jumping off point for future attacks. They were to be seen as an offensive rather than a defensive feature.[36]

Trenches and redoubts presented a particular problem for those planning manoeuvres, as they were forbidden in parks and woods or sites of archaeological interest. Digging into the turf of Salisbury Plain was restricted, so little experience of constructing earthworks was gained until the coming of war. The restrictions on the use of woods added a frustrating note of artificiality as this was just the sort of terrain which would have been useful for defensive positions. The 1897 Act also forbade 'interference with earthworks, ruins, or other remains of antiquarian or historical interest, or with any picturesque or valuable timber, or other natural features of exceptional interest or beauty.' Writing after the war, Lieutenant-General Snow was of the

35 Maj. Gen. E. Altham, *The Principles of War*, p.286.
36 S. Bull, *Trench: A History of Trench Warfare on the Western Front*, p.38.

opinion that 'the restrictions placed on manoeuvres in England so far as the use of the ground was concerned prevented the study of the use of ground in peace time on any large scale and had even taught us false lessons.'[37]

Detailed instructions were issued before each set of manoeuvres listing those areas which were to be out of bounds, such as houses, schools, parks and playing fields ('mounted troops and wagons are not to cross golf greens').[38] Such restrictions, unthinkable in a militarised society like that of Imperial Germany, were not unique to Britain. Before the 1907 Dutch manoeuvres, the mayor of Rheden in the eastern Netherlands wrote to the General Staff to remind them that officers were not to ride their horses on cycle paths along the highway, and that fines would result if this happened.[39]

For the 1909 manoeuvres, which took place around Lambourn, a celebrated horse training centre, there were added restrictions on crossing training gallops. Colonel Edward Altham was not alone in complaining about the anomalous situation whereby the Trespass Law was used to impede troops training over the land whereas 'any Tom, Dick or Harry who likes to spend two guineas in hiring a horse to follow hounds can ride over any man's land, and break down his fences, ride over his crops, and penetrate his covers, and so on.' Hastily reassuring his audience at the RUSI that he was not against hunting, Altham went on to argue that it would be easy to pass an Act of Parliament abolishing the Law of Trespass, so far as it concerned the exercising of troops on duty, with the taxpayer footing the bill when actual damage was done.[40]

There were also severe restrictions on the troops' use of farmland: 'Troops are not to pass through uncut corn or over clover seeds. Guns should not be fired so close to sainfoin [a forage crop] as to shake the seeds out.'[41] Some officers learnt to exploit these restrictions – one canny brigade commander is said to have created a powerful defensive position by covering his front with an out of bounds field of crops and his left flank with a barbed wire fence, which could not be touched.[42] This was a particular problem given that manoeuvres tended to take place at the same time of year as the harvest; compensation had to be paid to farmers whose crops were damaged, leading to speculation in 1903, when the harvest in Wiltshire and Hampshire was late and forced a week's postponement of the manoeuvres, that some farmers would have welcomed the compensation payments incurred by those troops with 'more than ordinary disregard for standing crops.'[43] This accords with the experience of a British

37 Lt. Gen. T. Snow, edited by Dan Snow and Mark Pottle, *The Confusion of Command: The War Memoirs of Lieutenant General Sir Thomas D'Oyly Snow, 1914-1915,* p.5.
38 Bowman and Connelly, op. cit., p.71.
39 H.P. van Tuyll van Serooskerken, *Small Countries in a Big Power World: The Belgian-Dutch Conflict at Versailles, 1919,* p.44.
40 Col. G. Ovens, 'Fighting in Enclosed Country with Some Notes from the Essex Manoeuvres', *Royal United Services Institution Journal,* p.543.
41 TNA WO 279/52, Report on Army Manoeuvres 1913.
42 Crawford, op. cit., p.13.
43 Ibid., p.16.

officer in the BAOR in the 1980s, who recalls a German farmer politely asking if they could 'accidentally' ram his ancient barn so that he could build a new one from the NATO compensation fund.

The harmful consequences of these restrictions, and the way that they could allow soldiers to slip into deeply ingrained bad habits, can be seen from an episode during the opening days of the Great War recounted by Lieutenant Rowland Towell of 41st Battery, RFA. His battery was moving up a hill when it came under German fire.

> It looked nasty and more shells arrived just in front on the road. I was ordered to wheel to the left across the field behind the hill. However, the lead driver of the leading gun went on obstinately up the road. The Battery Sergeant-Major galloped after him, cursed him for not following my order and asked him why he had not done so, when the man unexpectedly and in astonishment said, 'What, across that wheat?' The fruits of our careful peacetime training![44]

Army manoeuvres attracted large crowds of spectators, an unsurprising development given the arrival in quiet, rural areas of large numbers of troops, horses and guns, along with trucks, motorcycles and military bands. Local schools were often closed for the day when the action was nearby, and employers complained of high rates of absenteeism. After one set of manoeuvres, a landowner submitted a compensation claim for one day's wages for his two farm hands. When questioned about the army's responsibility, he replied that the men had stopped work to visit with the soldiers.[45]

Excitement was particularly intense if a member of the Royal Family was involved. King George V visited the 1912 and 1913 manoeuvres, while the Duke of Connaught, the third son of Queen Victoria, was a regular attender in his capacity as Inspector-General of the Forces between 1904 and 1907. Another attraction was the presence after 1909 of aircraft, as this was for many people their first chance to see an airship or aeroplane. In 1912, The *South West Suffolk Echo* observed that the inhabitants of Cowlinge had 'an exceptional opportunity of thoroughly inspecting these strange machines'.[46] The log book of Wolverton Infants School records for 23 September 1913: 'Army airship passed over in the morning. Children had good view from playground. 75 absent — nearly all taken by parents to Stony Stratford to watch the soldiers.'[47]

Spectators could cause problems for the army; one complaint was that members of the public supplied information, such as troop positions and movements, to participants, who then found it hard to ignore knowledge they should not have possessed. An additional problem was that the crowds could be so great as to impede the troops.

44 R. van Emden, *The Soldier's War: The Great War Through Veterans' Eyes*, pp. 43-44.
45 G.K. Scott-Moncrieff, 'Land for Military Training. An Appeal and a Suggestion,' *Blackwood's Magazine*, p.159.
46 *The South West Suffolk Echo*, 21 September 1912.
47 *The Story of Wolverton & New Bradwell 1913-1918*, http://www.mkheritage.co.uk/la/DaysofPride (accessed 30/8/2015).

This was especially true of the 1904 manoeuvres, and the extraordinary circumstances which saw troops being disembarked on a beach packed with holidaymakers. At the 1909 Conference of the General Staff, the Director of Military Training, Brigadier-General Murray, lamented the lack of control over spectators, which had reached its nadir at the previous year's cavalry manoeuvres on Salisbury Plain, when brigades moving concealed in low ground had been given away by the crowds of spectators moving parallel with them on the tops of hills. He noted that at German manoeuvres, where few such problems with spectators were encountered, cavalry NCOs were attached to the police to assist in crowd control. Murray did accept, however, that the contrast might be due to the German people's greater understanding of military affairs and their general discipline, whereas for the British, 'it is a free country, and they think they can do as they like.'[48]

There was a particular problem at the 1912 manoeuvres with heavy traffic. Murray, in his capacity as chief compensation officer, observed that the excellent roads and proximity to London had meant that large numbers of people had followed the manoeuvres in motor cars, often leading to congestion and getting in the way of troop movements, especially in the climactic battle on the final day. Furthermore, troops marching along the roads were often hampered by thoughtless spectators in motor cars, who passed at excessive speed and covered the men with dust. The following year The *Hemel Hempstead Gazette* of 20 September carried an advertisement warning motorists that roads in the immediate vicinity of the troops could be closed at short notice to wheeled traffic in accordance with the powers granted to the War Office under the 1897 Act. It went on to warn that 'persons desirous of seeing the Manoeuvres are advised, therefore, to come either on foot, on horseback, or on bicycles, for in motor cars or carriages they will be unable to get near.'[49]

Interactions between troops and members of the public could be fractious; a group of 80 Lancers on the 1904 manoeuvres reached the Wivenhoe ferry when the tide was at its lowest and forded the river Colne, only to find that the lady at the toll-gate on the other side demanded a penny per head for them to pass. The officer in command refused to pay, forcing the Lancers to turn back, re-ford the river, retreat along to Fingringhoe and use a ferry 'where there was no feminine or financial obstruction'. Needless to say, the local press made the most of the troops' discomfiture, employing the headline 'Eighty Lancers held at bay by a woman'.[50]

However, there were also benefits in manoeuvres taking place in full sight of the public. Given that many of its soldiers were either serving abroad or located in one of a few garrison towns, Britain's small professional army was remote from the general

48 TNA WO 279/25, Report of a Conference of General Staff officers at the Royal Military College January 1909, pp.58-59.

49 J. Sunderland and M. Webb, *All the Business of War: the British Army Exercises of 1913, the British Expeditionary Force and the Great War*, pp. 22-23.

50 *The Essex Newsman,* 10 September 1904.

Men of the Queen's Own Cameron Highlanders in Streetly End during the 1912 Manoeuvres. (© Hildersham History Recorders)

population to an extent that was not the case in Germany or France.[51] Some newspapers therefore argued that manoeuvres had an important role to play in creating a bond between army and society.[52] The Army was well aware of the public relations value of manoeuvres, which explains its willingness for them to be filmed – the *British Army Film,* made by Keith, Prowse and Company and released in January 1914 featured shots of an airship in action and troops on manoeuvres, while the Warwick Trading Company filmed the 1912 manoeuvres for its *Warwick Bioscope Chronicle* newsreel series, opening with shots of the King arriving to inspect his troops.

An army which had seen its reputation suffer in South Africa, and whose standing had been affected by periodic scandals was gratified and mildly surprised to find itself greeted with great enthusiasm during manoeuvres. The *South West Suffolk Echo* declared in September 1912 that 'it was a most moving sight to watch the inhabitants entertain "Tommy" with hot tea, coffee or cocoa, as well as eatables'. The same newspaper went on to deplore stories of local tradesmen putting up their prices when the troops arrived. It was 'not pleasant to reflect that there are people who call themselves Englishmen willing to deliberately swindle – for it is nothing else – our defenders'. Such people needed to be 'exterminated by force of public opinion'.[53] As Fanny Wale,

51 Bowman and Connelly, op. cit., p.147.
52 *The Pall Mall Gazette,* 19 September 1912, *The Morning Post,* 3 October 1913.
53 *The South West Suffolk Echo*, 21 September 1912.

of Little Shelford, five miles south of Cambridge, recalled many years later, 'The inhabitants of the Shelfords lived out of doors watching the Manoeuvres while they lasted, and felt dull when they were over', reflecting a widespread feeling that for many locals this might be the most exciting thing that ever happened in their village.[54] The following year a Buckinghamshire newspaper recorded even greater levels of enthusiasm:

> Wolverton practically lost its identity on Friday night, and looked more like a garrison town on the occasion of a Bank Holiday crush. Swarms of people flocked into the town from the neighbourhood, and the scene was a very animated one, and it was early evident, as the people began to mass along the thoroughfares, that the Gordon Highlanders were in for a great reception.[55]

Army manoeuvres were always attended by a number of military attachés from foreign powers, and these were as likely to be from neutral countries and future enemies as from allies. Thus the German military attaché, Major Martin Renner, attended the 1913 British manoeuvres less than a year before the outbreak of war, and within five years of Major Mustafa Kemal (later Ataturk) attending the 1910 French manoeuvres in Picardy, he was fighting against France and its allies in the Dardanelles.

Whether from allies, enemies or neutral countries, all had to be looked after in a suitable manner. They were put up in a comfortable local hotel (those in 1912 stayed at the Rutland Arms in Newmarket) and invited to a round of dinners and social events. For the 1904 manoeuvres in East Essex, the attachés stayed in the George Hotel in Colchester and were transported daily in a fleet of privately owned cars specially lent for the occasion. At the 1903 manoeuvres, there were 15 foreign attachés (four from Britain's ally Japan, three German, three French, two from the United States and one each from Italy, Spain and Portugal). The 1909 manoeuvres attracted observers from Austria-Hungary, Belgium, Bolivia, Brazil, Chile, China, Denmark, France, Germany, Italy, Japan, the Netherlands, Norway, Russia, Spain, Sweden, Switzerland, Turkey, and the United States. By 1912 the number of attachés attending manoeuvres had swollen to 32, with an additional seven Canadian and four French officers observing.

Looking after such a large party of foreign officers was an important responsibility, and one often assigned to a relatively junior officer keen to prove himself. Lieutenant-Colonel Aylmer Haldane, charged with looking after the foreign attachés at the 1903 manoeuvres in Wiltshire, found it a testing and time-consuming task. As well as escorting them to the manoeuvres themselves he had to look after their accommodation and organise a hectic social calendar; on the opening day, Haldane accompanied

54 Fanny Wale, *A Record of Shelford Parva* (1920), p.43. www.littleshelfordhistory.co.uk (accessed 30/8/2015).
55 *The Wolverton Express*, 5 September 1913.

Military attachés at 1910 French manoeuvres in Picardy – Mustafa Kemal stands ninth from the left. (Alamy)

Foreign attachés watching the action at Horseheath, September 1912. (Saanich Archives)

his charges to meet the Duke of Connaught, Umpire-in-Chief, at noon, before heading with them after lunch to Bowood House, seat of Lord Lansdowne, and then to Corsham Park, seat of Lord Methuen, for tea. After the manoeuvres ended, Haldane faced his most difficult task, settling the bill at the Castle and Ball Hotel in Marlborough: 'Had to fight with the hotel folk as they tried to stick me but failed. Total cost for 18 officers is only about £300 and in 1898 for 8 officers it was £250. Having Nice with me prevented any extravagance or swindling in the wine which I took from the A and N stores.' Nice was the wine butler at the Naval & Military Club (the 'In and Out') and assisted Haldane in keeping the spending on the attachés down to £285, even though the War Office had allotted £400 for the task.

What Haldane found most worthy of comment was that his charges had only consumed three dozen bottles of champagne in their five days in Wiltshire. Haldane had had an additional annoyance to cope with in that, having been appointed to the role, he found a senior officer, Colonel Thomas Pilcher, put in over him at the last moment. The attachés had been told not to bring horses with them to Wiltshire and nine motor cars had been allocated from the Volunteer Motor Corps for their use, but Pilcher proceeded to try to sell three horses to the German observers, Count von der Groben, Captain von Poseck ('a very nice fellow') and Count von der Schulenberg. Haldane told them to refuse if they were unhappy with their mounts 'as I think it very discreditable for an officer in charge of foreigners to be horse-coping'.[56]

Army manoeuvres took place in September as the culmination of what was known as 'The Collective Training Period'. The annual round of training began with individual instruction in the winter; squadron, company and battery training in the spring; regimental, battalion and brigade training in the summer, followed by divisional training, all building up to the army manoeuvres in early autumn. Training was referred to either as 'exercises' (carried out according to a detailed plan, with a specific objective) or 'manoeuvres' (where commanders had more freedom of action), though the two terms were often used interchangeably. One more term which requires definition at this point is 'Staff Ride.' A staff ride was a method pioneered by the Prussian General Staff to conduct officers over past campaigns or battlefields to study the decisions taken and the alternatives open to the commanders. It was originally carried out, as the name suggests, on horseback, but increasingly involved motor cars as well. Staff rides were increasingly used in the decade before the outbreak of war, with Douglas Haig a particularly enthusiastic proponent.

How did these manoeuvres actually work? Both sides were issued with the 'general idea', a document providing general information such as the context of the campaign and the frontiers of the countries involved, as these would be matters of common knowledge, whereas the 'special idea' for each side comprised the instructions upon

56 Aylmer Haldane, *Diary 1875 to 1913*, 13-19 September 1903. Pilcher's career was beset by controversy. Commanding a division, he was ultimately sent home from the Somme in disgrace in July 1916.

which it was to act. Since the two sides had identical uniforms, one wore a distinguishing mark, such as a white cap band. Blank ammunition was used and large numbers of umpires, wearing white arm bands, were on hand to supervise and to adjudicate on the result of combats, basing their decisions on a combination of factors including the numbers and type of troops involved, the terrain and the success with which units put the prevailing tactical doctrine into practice. The umpires ordered screens to be erected to represent casualties – these were small yellow cloth screens, carried by every half-company and used at the umpires' discretion when opposing forces approached within 800 yards of each other. If the umpire ordered that one screen be raised, that told the commander that he was receiving casualties and needed to check the pace of his advance. The second and third screens were raised to signify increasingly effective fire. If the fourth and final screen was raised, the commander had to change the disposition of his force or risk losing it entirely.[57]

Before the 1912 manoeuvres in East Anglia, *The Cambridge Daily News* explained to its readers the role of the umpires in the following terms: 'The duty of the Umpire Staff is to convey such information to the troops as would be imposed on them in war by the enemy's weapons.'[58] The challenge for umpires was to give detailed and specific descriptions to a commander without revealing facts about the enemy that he would not normally possess in a real battle. An umpire could only give a rationale for his decision if it did not betray the other side.

The numbers of umpires increased as manoeuvres developed in scale and complexity. A total of 108 umpires officiated at the 1903 manoeuvres, whereas at the 1912 manoeuvres there were 82 for the Red army alone, and even the Irish Command Manoeuvres of 1910, which were on a much smaller scale, employed an umpiring staff of 75.[59] Great reliance was placed on the competence and expertise of the umpiring staff, 'which represents the bullets',[60] and there were regular complaints in official reports about a lack of consistency in umpiring decisions. Smith-Dorrien was particularly critical of one decision at the 1910 manoeuvres, when the Red Mounted Brigade was adjudged by the umpires to have been 'annihilated' following a charge by the Blue Cavalry Division, commanded by Fanshawe, but was then allowed to resume operations after a half hour rest. This seriously undermined the manoeuvres and failed to reward Fanshawe's success. As well as being discouraging for the victorious Blue cavalry, 'there is a danger in teaching false lessons.'[61] The instructions for these manoeuvres made clear the need to avoid unrealistic outcomes – 'It must be distinctly understood that it is the duty of battalion and company commanders, as well as of the umpire staff, to prevent impossible situations arising.'[62]

57 G.T. Forestier-Walker, 'Umpiring', *Aldershot Military Society,* p.5.
58 *The Cambridge Daily News*, Saturday 7 September 1912.
59 TNA WO279/40, Report on Irish Command Manoeuvres 1910.
60 Grierson, op. cit., p.593.
61 TNA WO 279/39, Report on Army Manoeuvres 1910, p.53.
62 Ibid.

The 1907 manoeuvres on Salisbury Plain – Lord Methuen, Chief Umpire, receiving information from his staff. (*Black and White*, 7 September 1907, author's collection)

One cavalry officer, Douglas Haig, complained to General Sir Edward Hutton of an anti-cavalry bias during the 1903 manoeuvres: 'The umpires attach too much value to the killing power of the rifle, and constantly give faulty decisions, and force the cavalry to adopt a passive role.'[63] Smith-Dorrien was of the opinion that a perception, not always justified, of partisanship in umpiring decisions was unavoidable, especially when less experienced umpires were 'embedded' with a particular unit – 'It was asking too much of a junior officer to give an adverse decision in the case of a unit with which he was living, and whose commanding officer he had to meet at dinner nightly.'[64] A lengthy discussion of the training of umpires at the 1913 Staff Conference, chaired by Sir John French, ultimately rejected the idea of employing the system used by the French Army, in which umpires were attached to areas rather than units. That, according to Colonel Hunter-Weston, had been tried before and failed.[65]

Colonel Forestier-Walker, addressing the Aldershot Military Society in April 1911, warned prospective umpires not to worry about popularity:

> Don't expect that your decision will fill everybody with delight. From longish experience, I am inclined to think that there are only two occasions on which you are entitled to infer from the demeanour of the combatants that your decision has been a right one. One of these occasions is when neither side is pleased with

63 Stephen Bull, *Trench,* p.124.
64 TNA WO 279/47 Report on Army Manoeuvres, 1912 p.135.
65 TNA WO 279/48 Report of a Conference of General Staff Officers at the Royal Military College 1913, p.61.

> the decision. You may safely conclude that when this occurs fairly often, you are doing rather well. The other occasion is when both sides express approval of the decision. I have never yet met this case.[66]

Feelings between umpires and those engaged in manoeuvres could run high; at the 1903 Indian Army Manoeuvres a senior officer approached General Kitchener (Lord Kitchener's brother) to complain that an umpire had called him 'a bloody fool'. Walter Kitchener, either mishearing or pretending not to hear the exact words of the complaint, replied 'Quite right, quite right, an umpire's decision cannot be questioned; it must be accepted as correct by everyone', and the offended officer walked sadly away.[67]

Although there was an overhaul of the umpiring system between 1910 and 1911, concerns persisted over 'the selection of unsuitable officers out of touch with the training of regular troops as senior umpires.'[68] Battalion commanders were reluctant to release their best men for umpiring duties, while those who did volunteer to do the job were often doing so either in the hope of a cushy interlude or in order to get their own back on commanding officers with whom they had previously fallen out.[69] The CIGS, Sir Neville Lyttelton, considered the possibility of a permanent umpiring staff in 1908 but raised a predictable objection: 'Want of money was the chief difficulty.'[70] There was a persistent problem with the tendency of umpires to declare an action decisive without waiting to see how an extended fire-fight developed, often into a stalemate.[71] Sir Ian Hamilton was of the opinion that responsibility for the unsatisfactory conclusion to the 1912 manoeuvres lay less with the commanders than the chief umpires: 'They should have wider powers and should be the best men.'[72]

Because of a lack of blank adaptors, machine guns were silent at manoeuvres, and so they tended to be ignored by the troops.[73] Rawlinson found it dispiriting that his brigade trained hard with machine guns only to find that at the end of training they did not get credit from the umpires for effective fire. 'That is because they are hidden, and when they do not fire rapidly as they would in war, they got (sic) overlooked and no credit given to them.'[74] It could also be difficult for umpires to respond rapidly

66 Forestier-Walker, *Umpiring*, p.16.
67 Gen. Sir George Barrow, *The Fire of Life*, p.87.
68 TNA WO 27/508, Annual Report of the Inspector-General of the Home Forces for 1912, p.12.
69 Forester-Walker, 'Umpiring', p.2.
70 TNA WO 279/18, Report of a Conference of General Staff Officers at the Royal Military College, January 1908, p.25.
71 M. Connelly, Lieutenant-General Sir James Grierson in S. Jones, *Stemming the Tide*, p.144.
72 A.J.A. Morris, *Reporting the First World War: Charles Repington, The Times and the Great War*, p.161.
73 Bowman and Connelly, op. cit., p.92.
74 TNA WO 279/25, Report of a Conference of General Staff officers at the Royal Military College, January 1909, p.67.

to long-range artillery fire. Cavalry charges were especially difficult to regulate – in 1909 things got out of hand when scouts from a composite regiment of the Household Cavalry sighted the enemy (1st Cavalry Brigade) and both formations charged simultaneously. The Household Cavalry smashed into the ranks of the 1st Cavalry Brigade and two men from a Lancer regiment were killed and several injured.[75] Although manoeuvres were intended to be bloodless, wounds and sometimes even deaths did occur. There were seven fatalities associated with the 1912 British manoeuvres, and the Austro-Hungarian manoeuvres taking place the previous week resulted in eight men killed, with another six seriously injured.[76]

The workings of 'mimic warfare' might seem absurd, and certainly *Punch* had great fun with some of the more comical elements – the destruction of bridges was accomplished by posting a notice, signed by an umpire, declaring 'This bridge is destroyed' and detailing the width of the gap made and probable time needed for its repair[77]– but it is hard to see how they could have been avoided. As the Inspector-General of Cavalry, Major-General Allenby, remarked in 1912, 'All operations during peace are more or less unnatural, and more or less affected by conditions which will not prevail during war.'[78]

A further factor introducing a note of unreality was the practice of calling a cease fire at a certain point in proceedings. Operations were ceased at 3:30 p.m. on the third day of the 1903 Army Manoeuvres, with the two sides only resuming their positions at 10:30 a.m. the next day. In 1904 a twenty-four hour armistice was called on the second day to allow exhausted troops to recover from the landings made the day before. This approach, which hardly prepared troops for the relentless pace of real warfare, reflected the tendency of some commanders to see manoeuvres as a series of field days where the two sides were safe from enemy attack after 'close time' each day. Others, like Smith-Dorrien, argued that they should be seen as 'continuous operations in face of an enemy', and by the end of the decade manoeuvres were allowed to proceed uninterrupted until their conclusion.[79] When this was not done, William Robertson warned, 'the so-called manoeuvres did not give a true picture of what occurs in war, or anything like it, and might more appropriately have been called battle-drill.'[80] At the 1908 manoeuvres a cavalry brigade bivouacked at the end of the first day's operations, taking little or no precaution for security and unaware of the presence of an enemy infantry brigade within a mile or two. At dawn the next morning the cavalry were surrounded and the entire brigade taken prisoner. An umpire referred

75 L. James, *Imperial Warrior: The Life and Times of Field Marshal Viscount Allenby 1861-1936*, p.49.
76 *The Nelson Colonist*, 16 September 1912.
77 TNA WO 279/47, Manoeuvres 1912.
78 TNA WO 27/508, Report of the Inspector-General 1912, p.101.
79 Spiers, op. cit., p.88.
80 Robertson, op. cit., p.164.

**A WORLD OF SHAMS.**

*Officer (of Umpire Staff).* "HI, YOU THERE! YOU MUSTN'T CROSS HERE! CAN'T YOU SEE THE NOTICE? THIS BRIDGE IS SUPPOSED TO BE DESTROYED."
*Subaltern (cheerfully).* "OH, THAT'S ALL RIGHT! WE'RE SUPPOSED TO BE SWIMMING ACROSS."

A World of Shams, a cartoon parodying the rules governing manoeuvres. (*Punch*, 27 April 1910)

to Smith-Dorrien for a ruling on what should be done, and was informed that the cavalry were to be placed out of action until the end of the manoeuvres.[81]

As Commander at Aldershot between 1907 and 1912, Smith-Dorrien made determined efforts to increase the realism and hence the value of manoeuvres and exercises. Greater attention was paid to the quality of umpiring, to supply and transport problems and to co-operation between the different arms. These changes mirrored developments on the continent, where commanders sought to make manoeuvres a closer approximation of real warfare. Just as von Moltke in Germany attempted to introduce greater realism after his appointment in 1905, Conrad von Hotzendorf, Chief of Staff of the Austro-Hungarian Army between November 1906 and November 1911, ended the practice of working to a script, with manoeuvres culminating in a grand cavalry charge. He also put a stop to the practice of manoeuvres ending each evening and resuming with a new start line the following day. Henceforth the Dual Monarchy's generals would fight continuous campaigns of at least three days with the emphasis on encounter battles and the exercise of initiative and improvisation from commanders.

81 Marquess of Anglesey, *A History of British Cavalry 1816-1919 Volume IV 1899-1913*, p.420.

Observers from France and Italy were impressed by the change in approach evident in Conrad's 1907 and 1908 manoeuvres. Conrad also made far more use of staff rides than his predecessor, and made them more rigorous and relevant. On the 1908 General Staff Ride on the Upper Isonzo on the border with Italy, Conrad produced a plan which would eventually be employed against the Italians at Caporetto in 1917.[82]

Smith-Dorrien was widely recognised as having wrought considerable improvements to the Aldershot Command. Following the 1908 manoeuvres, *The Times* concluded that 'The admitted success of the manoeuvres must be highly gratifying to Sir H.L. Smith-Dorrien. The troops at Aldershot have received a systematic and progressive training, the results of which greatly impressed all who had the pleasure of seeing them.'[83] Even so, the degree to which manoeuvres were a realistic representation of the conditions of modern warfare was much discussed at the time. *The Manchester Guardian* stressed the difficulty of drawing meaningful lessons from exercises and manoeuvres, especially in the combat phase: 'Once troops become closely engaged in mimic warfare the operations cease to become either instructive or – to the soldier – realistic, and nothing can be gained by continuing the exercise, whatever its nature may be.'[84] *The Times* was unimpressed by the 1909 manoeuvres:

> The final position at which the issue was decided happily coincided with the place where the march past was to take place… it cannot be said that these rapid operations bore much resemblance to what might be expected in war in similar circumstances, and the unreliability of such a morning's work, in the opinion of many, exercises a dispiriting influence on the minds of those who take their training (seriously).[85]

*The Times* tended to be more sober in its description of events than the popular press, which felt no such inhibitions; *The Daily Mail*, which labelled the 1898 manoeuvres 'A Very Sham Fight', asked its readers: 'Why, it may well be asked, could not such a programme have been carried out with tin soldiers? The military mind can give no assistance, for the same reflection will be present to his mind.' The manoeuvres had cost the country £25,000 'without anybody – soldiers or generals – being one penny the better'.[86]

Such criticisms of the unreality of 'mimic warfare' were often based on a misunderstanding of the primary purpose of these manoeuvres. Speaking at the inaugural General Staff conference at Camberley in January 1908, the Director of Staff Duties,

82 L. Sondhaus, *Franz Conrad von Hötzendorf: Architect of the Apocalypse*, p.93, 94.
83 *The Times*, 21 September 1908.
84 *The Manchester Guardian*, 21 Sept 1912.
85 Crawford, op. cit., p.19.
86 *The Daily Mail*, 5 September 1898.

*First Territorial.* "Well, what do you think of our manœuvres, Bill?"
*Second Territorial (hitherto unacquainted with field-days).* "Thank 'evin we've got a Nivy!"

A *Punch* cartoon conveying reflecting contemporary views of army manoeuvres. (*Punch*, 20 May 1914)

Major-General Douglas Haig, identified six purposes beyond the obvious and basic one of preparing the army for war:

1. To give commanders experience of taking decisions and giving orders in circumstances resembling active service.
2. To give subordinates experience in executing orders.
3. To give the various arms and units practice in mutually supporting one another.
4. To test officers before promotion to a higher rank.
5. To give administrative services experience in feeding and supplying large numbers of troops.
6. To test tactical theories and new systems and equipment.[87]

They could be useful for practising supply and transport arrangements, and tactical lessons could be learnt, but increasingly the main aim, especially from 1906 onwards, was to prepare senior officers for the command roles they might soon assume in an

87 TNA WO 279/18, Report of a Conference of General Staff Officers at the Royal Military College, January 1908, p.20.

expeditionary force in a European War – as Grierson, now a Lieutenant-General, stated, 'Manoeuvres are a school for the leaders, in a less degree for the led.'[88] Colonel Altham agreed: 'They are there to train the generals and the brigadiers.'[89] This was especially important in a small volunteer army such as Britain's, where opportunities to command large bodies of men were few and far between. The Inspector-General of Forces, General Charles Douglas, worried that 'it is doubtful... whether we are training senior officers adequately in the art of command.'[90] When Sir John French went to France in August 1914 as commander of the BEF it was the first time he had commanded a force larger than a division other than in manoeuvres.

Manoeuvres could also provide invaluable experience for commanders at brigade and division level; James Edmonds, Chief of Staff to Major-General Thomas Snow's 4th Division, pointed out that the only practice which Snow had had in commanding a division in the field was during the three or four days' army manoeuvres each year.[91] Edward Bulfin, regarded by his corps and divisional commanders as performing extremely well between August and October 1914, had taken the unconventional staff administrative route to promotion, and had never commanded a battalion before his appointment to command a territorial brigade in July 1911, followed two years later by the prestigious command of 2nd Brigade of 1st Division. His experience of leading his brigade in the 1913 exercise was therefore his only taste of field command before the BEF set out for France.[92]

In an increasingly professional army which was approaching training and manoeuvres with a new seriousness, it was to be expected that the performance of officers in training and manoeuvres should affect their career prospects. As Inspector-General of Forces, Sir John French expressed his frustration that many senior officers did not possess a sufficient knowledge of the principles and practice of war on a large scale. French could be brutal in his observations in his annual reports, criticising one divisional commander for lacking strategical knowledge and handling his formation poorly on manoeuvres, and remarking of a Brigadier-General that he had not yet shown that he 'possessed sufficient knowledge of the art of war in its higher branches to fill his present post.'[93] French's successor Charles Douglas complained in 1912 that there were 'still too many officers in command... who are incompetent instructors, and inefficient commanders' and argued that the solution was 'the elimination of the unfit'.[94] The case of Major-General Henry Scobell suggests that this was sometimes done.

88 Grierson, op. cit., p.593.
89 Ovens, op. cit., p.542.
90 TNA WO27/508, Report of the Inspector-General 1912, p.8.
91 J. Edmonds, *Memoirs,* Chapter XXII, p.7.
92 M.S. LoCicero, 'A Tower of Strength': Brigadier-General Edward Bulfin in S. Jones (ed.) *Stemming the Tide*, p.214.
93 N. Evans, *From Drill to Doctrine,* p.302.
94 WO 27/508 Report of the Inspector-General of the Home Forces for 1912, pp.564-565.

Though a well-connected officer who had done well in South Africa, Scobell was criticised by two successive Brigade Majors, Hubert Gough and John Vaughan, for his slapdash approach to staff work and his resistance to new principles and training methods when in command of a cavalry brigade, and as Inspector-General of Cavalry from 1907 these traits served to damn him in French's eyes.[95] According to Vaughan, 'Goughie couldn't get on with him. Neither could I. The fact was that he had never been through the mill and didn't know essentials from non-essentials in training.'[96]

Scobell's nemesis came on Salisbury Plain on the first day of cavalry training in August 1908.[97] This was the first time the new Cavalry Division had been employed in manoeuvres; with four cavalry brigades drawn up in front of him, Scobell had no idea what to do with this mass of horsemen, and ordered them to advance 'in column of troops.' French adjudged that the cavalry 'were worked too much like infantry', and this was enough to bring about Scobell's dismissal.[98] Gough was unsympathetic: 'No wonder French dismissed him.' Scobell's career was not ended – he returned to South Africa to take up the Cape Colony command, remaining there until his death in February 1912 – but a clear statement had been made about the High Command's determination to raise the standard of training.

In the small world of an army consisting of only six divisions in peace-time, it was impossible for generals to avoid encountering each other year after year at manoeuvres, whether as rival commanders, umpires or directors. Tim Travers paints a picture of a high command riven by feuds and personal rivalries during this period and argues that these structural problems would linger during the First World War.[99] Patronage seemed to count for more than ability in an army still feeling the effects of the rivalry between the Wolseley Ring ('Africans') and the Roberts Ring ('Indians').

According to Edmonds, Lyttelton as Chief of the General Staff between 1904 and 1908 tended to favour officers from the Rifle Brigade and his successor Nicholson (1908-1912) officers from the Indian Army, while French pushed forward cavalrymen such as Allenby, Gough and Byng.[100] French's hostility towards the infantryman Smith-Dorrien, which was exacerbated by his resentment of the changes made by Smith-Dorrien after he succeeded French in command of I Corps at Aldershot in 1907, was such that at the 1909 manoeuvres conference he summed up in favour of Sir Arthur Paget and against Smith-Dorrien 'to the extent of being unfair to him'.[101]

95 N. Evans, *op. cit, p.302.*
96 Maj.-Gen. J. Vaughan, *Cavalry and Sporting Memories*, p.110.
97 S. Badsey, *Doctrine and Reform in the British Cavalry 1880-1918*, p.200.
98 NA WO 275/510 Report of the Inspector-General of the Forces, 1908.
99 T. Travers, 'The Hidden Army: Structural Problems in the British Officer Corps, 1900-1918', *Journal of Contemporary History*, pp.523-544.
100 T. Travers, op. cit., p.536.
101 T. Travers, *The Killing Ground. The British Army, the Western Front and the Emergence of Modern Warfare 1900-1918*, p.15.

Smith-Dorrien's biographer even makes the contentious claim that French 'went to some pains to see that his former friend was denied the opportunity of gaining practical experience in the handling of large bodies of troops at the annual manoeuvres by forbidding him on specious grounds to exercise his right to command.'[102] This relates to the 1910 manoeuvres, when five days before the scheduled start, Smith-Dorrien was replaced by Plumer as Commander of the Red forces (his Aldershot command) and given the role of Umpire-in-Chief.

An officer's performance in manoeuvres could certainly have a serious impact on his future prospects. In his diary, Haig attributed the fact that Grierson was switched from Chief of Staff to command of 2nd Corps in 1914 to French's displeasure when Grierson disagreed with his instructions at the 1913 manoeuvres.[103] Similarly, Rawlinson confided to *his* diary that French's dissatisfaction with his performance at those same manoeuvres was the reason why he was not retained in command of the 3rd Division when war came in August 1914: 'To introduce a personal matter into his choice of command was in my opinion petty but I had nothing to do but submit with the best grace I could.'[104]

In the same vein, Edmonds told Liddell Hart in 1937 that Arthur Paget had failed to get command of III Corps in 1914 because of the row he had had with French at the 1913 manoeuvres, though he was able to procure the command for his fellow Guardsman, William Pulteney.[105] However, Pulteney's biographer is sceptical of this story, and there was a less sinister explanation, which is that at 63 Paget was considered too old for a field command. It is also possible that his hapless handling of the Curragh Incident a few months before counted against him.[106] It has even been claimed that Charles Briggs was passed over for Corps command during the war because Haig remembered the part he had played in Haig's discomfiture in the 1912 manoeuvres.[107] If true, this would fit in with Travers' speculation, again derived from Edmonds, that Haig got rid of Brigadier-General Reginald Oxley in July 1916 because they had been enemies at Staff College, though it is far more likely that it was the failure of Oxley's 24th Brigade at Contalmaison which prompted his divisional commander to request his removal.

All of this strongly suggests a high command which was riven by favouritism and personal rivalry. However, rivalries and feuds were not unique to the British army – the progress of the right wing of the Schlieffen Plan in August 1914 was undermined by the repeated failure of von Kluck to take orders from von Bulow, while the Russian catastrophe at Tannenberg was at least in part attributable to the enmity between

102 A.J. Smithers, *The Man who Disobeyed, Sir Horace Smith-Dorrien and his Enemies*, p.132.
103 Douglas Haig (ed. G. Sheffield and J. Bourne), *War Diaries and Letters 1914–1918*, 13 August 1914, p.58.
104 N. Gardner, *Trial by Fire: Command and the British Expeditionary Force in 1914,* p.15.
105 T. Travers, *The Hidden Army,* p.536.
106 A. Leask, *Putty – From Tel-el-Kebir to Cambrai,* p.185.
107 G. Mead, *The Good Soldier,* p.436.

Generals Rennenkampf and Samsonov, who were members of the two rival factions which characterised the pre-war Russian army.[108]

There is also a major problem with so much of the evidence used by Travers and other historians such as Nikolas Gardner to prove their point about the amateurish and highly personalised nature of the pre-war British officer corps in that it tends to be anecdotal, and based on accounts from Edmonds, an entertaining but gossipy and unreliable source, or from Liddell Hart, a military journalist with a clear agenda who based many of his judgements on conversations with Edmonds and others in the 1930s.[109] Typical of this anecdotal approach is the story related by Tim Travers of Sir Arthur Paget, when GOC Eastern Command that he was:

> so lax that when in 1911 he commanded one of the forces at the annual army manoeuvres, he did not actually attend the manoeuvres, but spent his time in London. Consequently, Aylmer Haldane, his BGGS, had to give details of the manoeuvres to Paget while on his train from London to Salisbury, and a summary of what had happened, so that Paget could put forward a narrative of his forces' activities and answer questions at the post-manoeuvre discussion.[110]

On the face of it, this story conveys an impression of Paget as hopelessly lazy and incompetent, and the Army as amateurish for tolerating such conduct. However, the story seems too good to be true – could the commander of one of the two sides in the annual manoeuvres, the biggest event in the peacetime army's calendar, really have been absent while they took place? The first clue that Travers has the story badly wrong lies in the fact that the September 1911 army manoeuvres never took place, being cancelled due to drought conditions in Wiltshire.

Recourse to Haldane's diaries in the National Library of Scotland reveals that what Travers describes as annual manoeuvres was in fact a staff ride which took place in May 1911 and which Haldane was to lead while Paget fulfilled official duties at Court. Haldane records on 11 May that he 'left by 8.40 train from Liverpool St. AP (Paget) was on it so could cram him with what was necessary for the conference.' Haldane went on to record that 'AP is wonderfully quick at picking up the threads for a conference, and I had not to utter a word beyond what I had told him in the train.'[111] The impression that Travers gives of Paget's competence is misleading, to say the least.

Further proof of the unreliability of so much of the evidence cited by Travers comes from Edmonds' allegation that Sir Henry Wilson had intrigued to ensure that he succeeded Rawlinson as Commandant of the Staff College in 1906 by using his

108 The old story that their enmity dated back to a physical altercation during the Russo-Japanese War has been debunked by D. Showalter, *Tannenberg: Clash of Empires*, p.134.
109 For a characteristically balanced and amusing judgement of Edmonds, see P. Griffith, *Battle Tactics of the Western Front: The British Army's Art of Attack 1916-1918*, pp.258-260.
110 T. Travers, *The Killing Ground*, p.27.
111 A. Haldane, *Diary 1875 to 1913*, 11 May 1909.

position as Deputy Director of Staff Duties to remove the two leading candidates, Lawson and Capper, to the Dublin Infantry Brigade and the Indian Staff College at Quetta respectively. As John Hussey has pointed out, many of the alleged facts do not fit, and Edmonds had a strong animus against Wilson, who was in any case the strongest candidate for the post.[112] Hussey concludes that 'This story… is worthless as evidence to prove anything about the structural defects of the old Army' and offers an invaluable warning for those seeking to base a case on the recollections of Edmonds and Liddell Hart:

> When next you see some hearsay quoted from these sources, ask for a little independent corroboration, and if you see this old story again in some new book, put a little question mark against it and look very hard indeed at the work in which it appears.[113]

Whatever their usefulness for the commanders, George Barrow, cavalry instructor at the Staff College, was of the opinion that the primary purpose of manoeuvres was to build team spirit:

> The four cavalry brigades moving over the plain on a summer's day made up a superb picture that bore no resemblance to war as we came to know war not long afterwards. Not that these manoeuvres were wasted: they had the inestimable value of uniting staffs, commanders and regimental officers in community of method and mutual understanding, and strengthening the bond of comradeship that was a marked feature in the old pre-war cavalry.[114]

Barrow's experience of the 1903 Indian Army manoeuvres, which 'passed off in much the same way as most manoeuvres', had made him profoundly cynical about the whole process: 'The troops marched hither and thither without knowing why; the junior ranks knowing little more, slightly bored.' Meanwhile, the senior officers displayed 'an intelligent interest in the lessons learnt and the faults to be corrected as set forth at the "pow-wows", the final conference and summing up, and the concluding pat on the back, for the great keenness shown by all ranks.'[115]

However much British army manoeuvres of this period may be criticised for being unrealistic, they were no more so than those of continental armies – his years in Berlin had taught Grierson that British exercises 'are far more like war than the ridiculous shows here,'[116] while Aylmer Haldane, who commanded a brigade in both 1912 and

112 J. Hussey, 'Appointing the Staff College Commandant 1906: A Case of Trickery, Negligence, or Due Consideration?' *The British Army Review*, pp. 99-106.
113 Ibid., p.104.
114 Barrow, *The Fire of Life*, p.118.
115 Ibid., p.87.
116 D. Macdiarmid, op. cit., p.138.

1913, concurred with the cabinet minister John Burns: 'We agreed that in many ways the great annual manoeuvres in Germany and France were unreal and did not represent actual war or anything like it, for it seemed that they were conducted to impress spectators, especially those who knew little about military matters.'[117] It has been argued that the scathing criticisms of the 1911 German manoeuvres made by Charles Repington in *The Times* were deliberately dismissive in order to undermine the aura of German military invincibility, but he was only repeating what many observers, Germans as well as foreigners, had said about the Imperial Manoeuvres.[118]

Horace Smith-Dorrien was shocked by what he saw at the 1905 German manoeuvres, recording in his diary for 4 September: 'A hammer-and-tongs battle with quite impossible situations, over in half an hour after infantry came in contact.[119]' George Barrow was equally unimpressed by what he saw of the French army: 'The manoeuvres were a grand spectacular show bearing no resemblance to modern warfare. This is no reflection on the French military authorities, for the same remark applies to the German and all other army manoeuvres, including our own.'[120] The Socialist leader Jean Jaurès complained in 1910 that French manoeuvres were used by ambitious generals to advance their careers rather than to prepare for war:

> The grand manoeuvres are nothing but a parade where military leaders hope to be noticed, not through good planning and organisation, but by the press and politicians. The point is not who best directs his forces to achieve precise goals, but who will have the most influential newspaper editor in his car... The best part of their strategy goes into press campaigns against their rivals, while battalions, regiments and brigades move in a void, without firm direction or goal.[121]

A comparison with the German Imperial Manoeuvres, held in September each year, is especially illuminating, given the general acceptance of the notion that the German Army was well ahead of the British in terms of preparedness and competence. German manoeuvres in the two decades before the outbreak of the First World War were deeply flawed. This was partly due to the innate conservatism and suspicion of change exhibited by the officer class, many of whom saw no reason to abandon the methods which had secured victory against the French in 1870-71. However, the main reason for the failings of the German manoeuvres in the pre-war period was the unpredictable and unhelpful contribution of Kaiser Wilhelm II. The 'All Highest' insisted on participating in the manoeuvres, personally taking a hand on the final day each year, leading a spectacular massed cavalry charge to clinch victory, and he was

117 Brig. Gen. Sir Aylmer Haldane, *A Soldier's Saga*, p.273.
118 Dorn Brose, op. cit., p.157.
119 C.R. Ballard, *Smith-Dorrien*, p.84.
120 Barrow, op. cit., p.117.
121 D. Porch, *The March to the Marne: The French Army 1871-1914*, p.178.

indulged by sycophantic courtiers and generals, who knew that the Kaiser always had to be on the winning side. Manoeuvre grounds were chosen purely on their suitability for cavalry action, and infantry were encouraged to attack strong defensive positions across open ground in a manner that would be suicidal in a real war. As early as 1894, General Colmar von der Goltz was complaining that the Chief of the General Staff, von Schlieffen, should not have permitted the Kaiser's theatrical charge with 60 cavalry squadrons – 'It is dangerous to practise something that must surely fail in war.'[122]

The 1903 manoeuvres marked the zenith of this approach – the commanders of the Guard Corps protested,' seeing in these things a playing at soldiering which does not correspond with the purpose of the army or seriousness of the internal situation, and which will lead to excessive mockery and criticism by agitators' (meaning the socialists), while the Bavarian military attaché Karl von Endres complained to the War Minister in Munich that the outdated tactics on show 'do harm to the already endangered reputation of the army with the people.'[123] A young cavalry officer from the United States, Lieutenant Frank R. McCoy, was shocked by what he saw during these manoeuvres, and in particular the lack of realism in their massed cavalry charge, which demonstrated that 'the Germans had not had the illuminating experience that the modern rifle shoots into one.'[124]

Things were no better the following year, when Wilhelm's court chamberlain, Count Robert von Zedlitz-Trützschler, observed of the 1904 manoeuvres in Mecklenburg-Schwerin: 'I am told that everybody, even the umpires, usually know which way the wind is blowing, and their chief anxiety is to act accordingly.' As a result, 'the manoeuvres of this year were just the same as last year's, and have not been an atom of use to either the troops or the General Staff.'[125] Colonel Alexander Godley, commander of the British Mounted Infantry school at Aldershot, noted that 'attacks were made in massed formations' and that on one day the Kaiser commanded one side, the next day the other, and each time his side had the upper hand. On the final day 'the Kaiser put himself at the head of the cavalry division, and on a white horse, and followed by a huge staff, charged serried masses of infantry and guns and put them to utter rout!'[126]

Things improved to a certain extent after 1905, when von Moltke made it a condition of his accepting the role of Chief of the General Staff that the Kaiser must withdraw from participating in manoeuvres.[127] For the next six years, there was a more professional approach, with the Kaiser mostly resisting the temptation to interfere and von Moltke leading a younger generation of serious-minded generals such as

122 Dorn Brose, op. cit., p.124.
123 Ibid., p.125.
124 Frank R. McCoy, Notes on the German Maneuvers, *Journal of the U.S. Cavalry Association* 14 (July 1903), p.27.
125 Count Robert von Zedlitz-Trützschler, *Twelve Years at the German Court*, p.88.
126 General Sir Alexander Godley, *Life of an Irish Soldier*, p.108.
127 A. Bucholz, *Moltke, Schlieffen and Prussian War Planning*, p.243.

Friedrich von Kleist and Friedrich von Bernhardi, determined to use manoeuvres to test new tactics and technology. However, this adjustment did not come easily to many officers in such a vast and socially conservative officer corps, and in his report on the 1911 autumn manoeuvres von Moltke was highly critical of some infantry units for attacking too quickly and having skirmish lines which were too tightly packed together.[128] Ominously, by 1912 the Kaiser had reverted to interfering, admonishing his generals after the 1913 manoeuvres for favouring flanking movements rather than frontal assaults.[129]

A similar process of regression was to be found in the armies of the Austro-Hungarian Empire, where the chief of staff, Conrad von Hötzendorf, had to cope not so much with his monarch as with the heir to the Habsburg throne, Franz Ferdinand. Conrad's dismissal in November 1911 enabled the Archduke to reverse the move towards greater realism and return to the formality and ceremonial favoured by Conrad's predecessors; this process began with the September 1912 manoeuvres in Hungary, at which Franz Ferdinand commanded and Conrad was a hapless observer while his less assertive successor Schemua acted as chief of staff. Even after Conrad was restored to the post in December of that year, he faced interference from the Archduke, newly emboldened by his appointment as General Inspector of the armed forces. Franz Ferdinand insisted on altering Conrad's plans for the 1913 manoeuvres, held in Chotowin in Bohemia, extending them by two days to incorporate a grand cavalry charge commanded by Franz Ferdinand; Conrad wrote to his mistress to complain bitterly that the Archduke 'has arranged the whole thing... as a military spectacle for his wife, his children, and the Bohemian nobility.' He dismissed the whole exercise as 'a spectacle for layman and children'.[130]

Nor was this problem unique to the armies of the Dual Alliance; King Willem III of the Netherlands, who ruled between 1849 and 1890, frequently tried to take command of manoeuvres and created chaos wherever he went.[131] In contrast, Wilhelm II's cousin George V attended manoeuvres purely in an observer's role and limited himself to a few encouraging comments at the final conference, as he did in 1912 and 1913.

The first set of British army manoeuvres after the conclusion of the Boer War in May 1902 saw two armies, each of approximately 20,000 men, take part in a four-day campaign in September 1903. The area proclaimed under the Manoeuvres Act formed a large quadrilateral between Banbury in the north to Windsor in the south east, down to Portsmouth to the south west and across to Bath in the west, but the important action took place to the north of Newbury, around Lambourn and Shefford (where the M4 motorway now cuts through the Downs). The Red army, made up of

128 Dorn Brose, op. cit., p.157.
129 Ibid., p.181.
130 Sondhaus, op. cit., pp.119, 131-132.
131 Dik van der Meulen, *Konig Willem III*, p.257.

the First Army Corps under General Sir John French, was successful in halting the advance of the Blue force, commanded by Field Marshal Evelyn Wood, who was moving east towards Reading from his base in Wiltshire. Two events were decisive in the Red success, the first when the Blue Fifth Division was caught in flank by both Red divisions as it moved north around Shefford and suffered heavy losses, and the second when a Blue cavalry charge down a valley east of the village of Fawley was a complete failure, thanks to fire from the Red mounted infantry and an infantry battalion and the guns of the Red cavalry.

The Commander in Chief, Roberts, as Director of Manoeuvres, was lavish in his praise of the infantry, who had shown the extent of the lessons they had already learnt from South Africa – 'they took cover well, and both in attack and defence showed the great improvement in the training of that arm as a result of our experiences in South Africa' – and less so when it came to the artillery. He had been disappointed to see the guns so exposed, as if nothing had been learnt from the experiences of the veldt: 'On the whole there was practically very little attempt at concealment, and in several cases batteries neglected infantry fire in a manner which would have caused them unnecessarily heavy loss.' However, his especial displeasure was reserved for the cavalry, which he felt had shown little progress since the Boer War: 'I am not satisfied that in these manoeuvres the cavalry have displayed quite as much independence or initiative as I had expected would have been the case after our experiences in South Africa.'[132] When reading these comments it is worth bearing in mind that Roberts was already engaged in a heated tactical debate over the future role of the cavalry in which he found himself on the other side to French and Haig, who at the risk of simplification might be termed the 'shock action' school as opposed to those like Roberts and the influential author Erskine Childers who stressed the cavalry's prime purpose as providing dismounted fire action.[133]

Generally, Roberts' perception of genuine progress made in the short period since the Boer War, at least from the infantry, was shared by press and generals alike. Grierson was in a good position to judge, having been an umpire at the unsatisfactory 1898 manoeuvres and served on Wood's staff in 1903, and he recorded in his diary that the infantry had been 'marvellous' and had manoeuvred in a way undreamt of five years earlier.[134] One American observer, Captain Lawrence Timpson of the New York National Guard, took away with him as his overriding impression the qualities of the British private soldier:

132 TNA WO 279/7, Report on Army Manoeuvres, 1903.
133 The polarisation between the two sides in the debate has masked the extent to which they agreed on several key issues. See S. Jones, op. cit., pp.170-196 and Bowman and Connelly, op. cit., pp.180-181.
134 Macdiarmid, op. cit. p.202.

At the end of the week's work, coming as it did after the preliminary manoeuvres, and weeks in camp under the most trying weather, he was as cheery and smart as though he had just come out of a barrack yard. He took a most intelligent interest in the manoeuvres, carrying out his part of them, the uninteresting detail, faithfully to the end. He is physically fit and joins the service for the love of it.[135]

135 Report of Captain Lawrence Timpson, Inspector of Small Arms Practice, First Regiment, on the British Army Maneuvers (1903), http://www.fold3.com/image/301245521 (accessed 30/8/2015).

2

# Invading Essex: The 1904 Manoeuvres

The 1904 manoeuvres, in which a force of 12,000 men, 42 guns and 2,700 horses was landed on the coast of Essex, were unique among those undertaken in this period in that they involved an amphibious operation. They were also unusual in taking place away from Salisbury Plain and the open countryside to which the army was accustomed in its training. The stimulus for this innovative approach came from the widespread fear of foreign invasion which exercised such a powerful influence on politicians and the British public alike at the time. The unfolding events of the Russo-Japanese War, and in particular the stunning Japanese attack on the Russian fleet at anchor at Port Arthur in February 1904, appeared to confirm the danger to Britain in the new age of modern steam fleets. St Loe Strachey, writing in *The Spectator,* of which he was the editor, declared that 'To those who are on the land and expecting an attack from the sea, the water becomes a place of dreadful mystery from which at any moment a sudden and unexpected blow may fall.'[1]

Strachey had previously advocated holding manoeuvres which could test the ability of the Auxiliaries to cope with an invasion close to London, and the War Office was well aware that France, Germany and Russia had all held amphibious manoeuvres in 1901.[2] Even more recently, the Austro-Hungarian army and navy had staged joint manoeuvres in September 1902, which assumed an enemy amphibious assault against the naval base at Pola.[3] The Secretary of State for War, Hugh Arnold-Forster, brought a scheme for combined manoeuvres to the first meeting he attended of the Committee for Imperial Defence on 18 November 1903. The following day, Admiral Sir John Fisher, shortly to be appointed First Sea Lord, informed Lord Esher that 'Instead of our military manoeuvres being on Salisbury Plain and its vicinity (ineffectually aping the vast Continental Armies!) we should be employing ourselves in joint naval

1 H. Moon, The Invasion of the United Kingdom: Public Controversy and Official Planning 1888-1918 (PhD thesis, London University, 1968), p.247.
2 Ibid., p248.
3 L. Sondhaus, *Franz Conrad von Hötzendorf: Architect of the Apocalypse*, p.65.

and military manoeuvres, embarking 50,000 men at Portsmouth and landing them at Milford Haven or Bantry Bay!'[4]

Arnold-Forster quickly became the driving force behind the scheme, arguing that 'We have had manoeuvres on and off for 20 years, and 3 years war in Africa, but have never yet tried the only problem which concerns us.'[5] He rapidly encountered opposition to the project from the Army. The Commander in Chief, Lord Roberts, was unconvinced, stating 'Before giving an opinion, I would like to talk the proposals over with some sailors, for I confess I am somewhat skeptical as to their being of much practical value.'[6] When the Army Board discussed the scheme on 15 December 1903, one officer warned that far from producing co-operation between the two services, it might lead to further friction between them. Arnold-Forster conceded that this was a danger, but suggested that the Admiralty should provide a Flag Officer and a squadron for the attacking force: 'Such an officer would, I have no doubt, enter thoroughly into the scheme, and would devote all his skill to outwitting and defeating his naval colleague with the defending force.'[7] It is important to note that at this stage the operation was conceived as testing the Navy's ability to prevent a landing as well as to enable one. At this stage in the planning, Arnold-Forster was interested in the landing itself rather than what happened subsequently: 'The military part of the defence is… not of great importance, as if the landing can be effected, the problem is thereby solved in favour of the invader.'[8] The scheme as ultimately put into practice would show major changes from this original intention.

Having won over the Army Board, which decided in favour of the scheme on 14 January 1904, Arnold-Forster now had to convince the Navy. Fisher had cheerily predicted 'banter' between the two arms – 'No doubt there will be good-natured chaff.'[9] However, what Arnold-Forster and the Army's Director of Military Training, Brigadier-General Frederick Stopford, actually encountered was an outright refusal to take part. Having met Prince Louis Battenberg, the Director of Naval Intelligence, Stopford reported that 'the Admiralty have now apparently given up the idea altogether.'[10] Arnold-Forster, not one to be daunted by finding himself in a small minority, pressed on with his plans for an operation in Essex involving 30,000 troops, but the lack of Admiralty co-operation at this stage of the planning meant that the project became a military one not very different from previous military exercises. In particular, the naval aspects of an invasion were neglected, with serious implications for the way the manoeuvres would eventually unfold.

4 A.J. Marder, *Fear God and Dread Nought Vol. I*, p.291.
5 Moon, op. cit., p.249.
6 Ibid.
7 Ibid., p.250.
8 Ibid.
9 Marder, op. cit., p.291.
10 Moon, op. cit., p.251.

Nevertheless, these were the first army manoeuvres in Britain ever to involve the Navy in co-operation with land forces. It was obvious that no expeditionary force could ever leave British soil until the Navy had secured command of the sea to ensure its freedom of passage, so the need for naval involvement in the manoeuvres was obvious. Some thought was given to having part of the Navy represent the naval power of the defence as well as another part playing the role of the fleet securing the passage of the expeditionary force, but practical difficulties became apparent when they considered how this was to be done – in order to gain any meaningful experience, the defenders' naval force would have to cover a long stretch of coast line, and a wide choice of landing places would have to be offered. The Military Manoeuvres Act of 1897 made it necessary to proclaim six months beforehand the districts chosen, and a district once proclaimed could not be proclaimed again for five years, thereby ruling out the whole coastline for future use for that period. Therefore, it was decided that the naval element should not involve the defenders; it was assumed that the attackers had obtained command of the sea, so the landing would not be opposed. Naval-military co-operation would have to 'be exhibited in the disembarkation of an expedition unopposed by naval force.'[11]

The Essex coast line between Clacton-on-Sea and Holland-on-Sea was chosen, an obvious contrast to previous locations for manoeuvres: 'Hitherto, manoeuvres have taken place at Aldershot, on Salisbury Plain, and on the Berkshire Downs, where the country is open and where very marked positions exist.' The Army had 'very rarely, or never been exercised in operations in a close country such as the County of Essex presents.'[12] This was much more typical of the English countryside and the places where an invasion would have to be resisted, which was thought to be a real danger in an atmosphere stirred by invasion literature such as *The Riddle of the Sands*, published the previous year.

Erskine Childers' novel, which features two men foiling a German invasion of England when they stumble across a fleet of barges being assembled on the German coast, took pride of place among an extraordinary number of novels (over 400), plays and articles of the period which played upon the British fear of invasion. Authors who contributed to the genre included Saki (*When William Came*, 1913), John Buchan (*The Thirty-Nine Steps*, 1915), H.G. Wells (*The War in the Air*, 1908, and, with a science fiction twist, *The War of the Worlds*, 1898) along with a host of lesser writers, while it was parodied in inimitable form by P.G. Wodehouse in *The Swoop! Or, How Clarence Saved England* (1909), in which as well as the Germans 'no fewer than eight other hostile armies had, by some remarkable coincidence, hit on that identical moment for launching their long-prepared blow.' While the Germans landed in Essex, the other landing spots for the Russians, Swiss, Chinese, Young Turks, Moroccan brigands, the Mad Mullah and his followers and the Prince of Monaco included well-known

11 TNA WO 279/8, Report on Army Manoeuvres 1904, p.15.
12 Ibid., p.18.

coastal resorts – Scarborough, Brighton, Margate.[13] Home defence enthusiasts such as Strachey seized on invasion literature, much of it written by serving officers in the Army and Navy, to support their case for increased military expenditure and the introduction of conscription.[14] In such a climate, an amphibious landing at Clacton-on-Sea did not seem such an outlandish proposition.

The War Office carried out a series of smaller exercises to prepare for the larger one. Brigadier-General F.H. Lake led a staff ride 'based on the supposed landing of a hostile force on an open beach, the establishment of a base of operations, and the subsequent advance inland.' An unopposed landing of heavy artillery on Whale Island was conducted in the presence of Field-Marshal Wood and War Office officials, and a landing of men and guns took place at Bournemouth, all between 19 and 22 April 1904.[15]

The value of the manoeuvres was undermined by a confusion, never satisfactorily resolved, over their fundamental purpose – was it to test the Army's ability to launch an invasion on a hostile coast or to resist an invasion of Britain? If the latter, the proviso that the landing was to be unopposed made the exercise almost worthless. Arnold-Forster had been in no doubt as to the purpose – 'We do not want to practise landing an Army on a foreign coast, but we do want to see whether a foreign Army can land on our coast.'[16] However, by the time the manoeuvres took place in September 1904, the emphasis had shifted to a more orthodox military problem. Even so, the problem of how to disembark a sizeable force of all arms on open beaches was still one worth addressing.

The 'General Idea', setting out the context of the manoeuvres, posited that Blue had landed on the coast of Sussex, driving back the Red forces; Red was now concentrating in the Redhill area to contest the further advance of Blue. The Chief Umpire was the Duke of Connaught, Inspector General of the Forces, with General Lord Methuen the Blue Senior Umpire and Major-General Laurence Oliphant (GOC Home District) the Red. The War Office's new Director of Military Operations, Major-General James Grierson, was one of thirty Blue umpires (the smaller Red force had sixteen). The whole set of manoeuvres was under the direction of Lieutenant-General the Hon. Sir Neville Lyttelton, the newly appointed Chief of the General Staff. The Directing Staff would be headed by Sir Frederick Stopford, and the Blue (invading) army would be commanded by Sir John French, fresh from his successes in South Africa. French still had some fondness for his old service, having spent two years as a Midshipman on HMS *Warrior* between 1868 and 1870; In a vote of thanks to Commander Ballard of the Naval Intelligence Department after a lecture on 'Naval

13 I.F. Clarke, *Voices Prophesying War: Future Wars 1763-3749*, p.127.
14 For the influence of invasion literature on military planning, see A.M. Matin, 'The Creativity of War Planners: Armed Forces Professionals and the pre-1914 British Invasion-Scare Literature', *English Literary History*, pp.801-831.
15 *The Times*, 20 April 1904.
16 Moon, op. cit., p.251.

and Military Co-operation in War' at Aldershot the previous year, he had welcomed the idea of joint manoeuvres: 'I sincerely hope that, in years to come, the annual naval and military manoeuvres will be carried out in conjunction, a matter we would welcome very much as being of great national service.'[17]

French's army, consisting of I Army Corps, contained two divisions, the 1st under Major-General Arthur Paget and the 2nd under Major-General Sir Bruce Hamilton, a total of 11,571 troops, with 61 guns and 2,701 horses.

Following the landing of his troops on a two mile stretch of beaches between Clacton and Holland-on-Sea, French was ordered to push inland, seize Colchester and head for Maldon, which would act as his base of operations. The seizure of Maldon was French's main objective and an urgent one, as it was thought to be a hazardous procedure to rely for one's line of supplies on an open beach – earlier in 1904 Fisher had insisted that 'no rational commander would rely on landing on an open beach. Hence, some sort of harbour must be used.'[18] It was therefore essential to seize a harbour with wharfage for heavy stores, and Maldon offered such a base, though the port was only suitable for craft of shallow draught. The stretch of coast between Clacton and Holland was deemed ideal because of its gently sloping beaches and a continuous bank of low cliffs offering protection to landing parties. The Clacton and Holland-on-Sea Coast Protection Scheme, completed in 2015 at a cost of £36 million, has done a remarkable job of combating coastal erosion and means that one can gain a much better sense than before of what the beaches on which the troops landed may have looked like in 1904.

The Red commander, Major-General Arthur Wynne (GOC 10th Division) had a force only half the size of Blue, made up of two brigades of infantry, a small force of Lancers, squadrons of Life Guards and Horse Guards, a squadron of Yeomanry, Engineers, and fifty cyclists. The Red force was only intended to delay Blue while awaiting reinforcements – typically, there were also budgetary considerations behind the size of the defending force, as 'it was necessary, for financial reasons, to limit expenditure as much as possible.'[19]

Ten transport ships were hired from British shipping companies, mostly from the Leyland Line and the Atlantic Transport Line, at an estimated cost of £68,000 (£43,600 for the hire of ships, fittings £14,600, coal £2,300, dock, pilotage and light dues £5,000) to transport the troops from Southampton to Clacton.[20] Of these ten steamships, which were on average of between 7,000 and 9,000 gross tonnes, four would see action as troop transports in the Great War: SS *Menominee* was involved at Gallipoli, transporting Australian troops from Alexandria to the Dardanelles,

17 Major the Hon. G. French, *The Life of Field-Marshal Sir John French, First Earl of Ypres*, p.137.
18 N.A. Lambert, *Sir John Fisher's Naval Revolution*, p.89.
19 TNA WO 279/8, Report on Army Manoeuvres 1904, p.18.
20 Ibid., p.43.

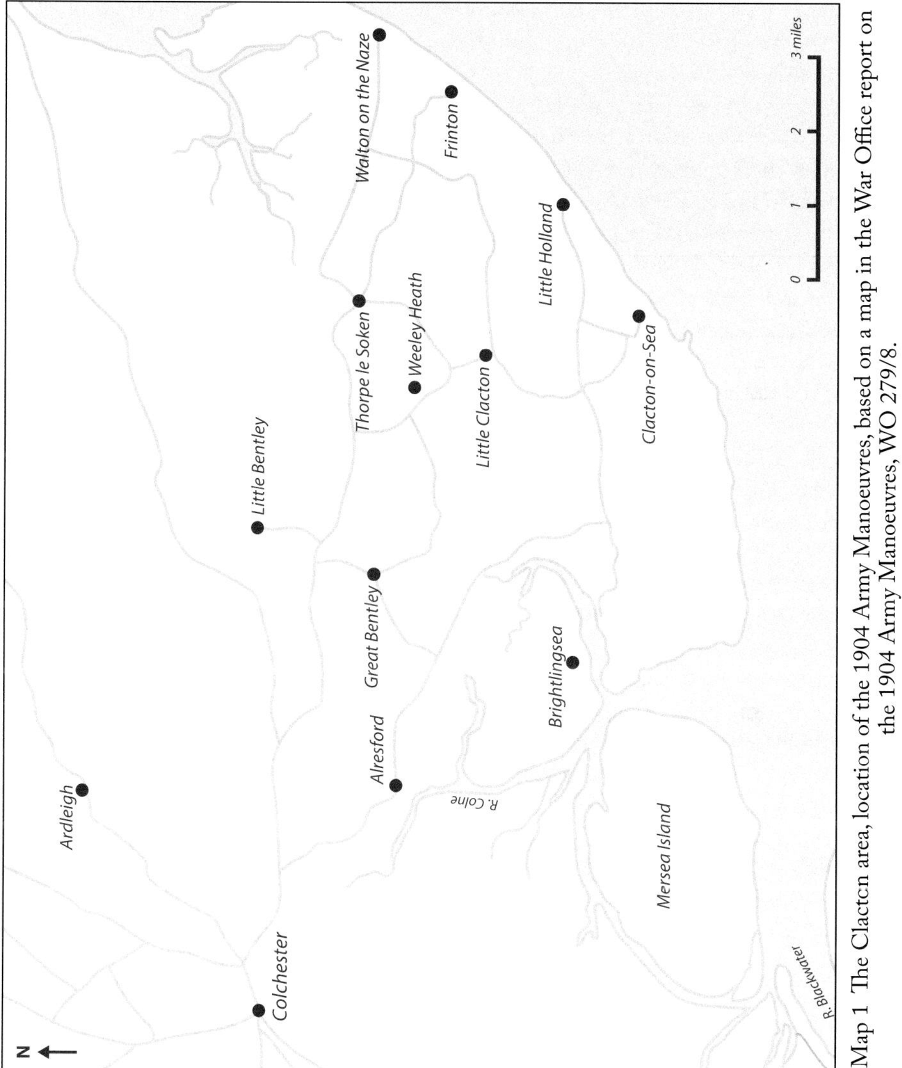


Map 1 The Clactcn area, location of the 1904 Army Manoeuvres, based on a map in the War Office report on the 1904 Army Manoeuvres, WO 279/8.

SS *Marquette* transported the 12th Light Horse to Gallipoli in August 1915 and sank with the loss of 167 lives in October of that year en route to Salonika, and her sister SS *Manitou* was used to transport 29th Division, being attacked on the way to Mudros by a Turkish torpedo boat on 16 April 1915, with the loss of 47 soldiers and 7 crewmen. SS *Atlantian* was also involved in landing troops at Gallipoli.[21] The assumption must be that some of the crews at Gallipoli had also taken part in the Clacton landings a decade earlier.

The transports were to convey 559 officers, 11,139 men, 2,704 horses, 61 guns, 315 vehicles, four motors and 108 bicycles from Spithead to Clacton. They were accompanied on their journey by the Cruiser Squadron under its commander Rear-Admiral Sir Wilmot H. Fawkes; this squadron consisted of six brand new, state-of-the-art armoured cruisers, all launched in the previous three years, two Drake-class armoured cruisers (*Drake* and Fawkes' flagship *Good Hope*), each of 14,100 tonnes, with a crew of 900, and four Monmouth-class armoured cruisers (*Berwick, Donegal, Kent and Monmouth*), each 9,800 tonnes, with a crew of between 675 and 720. All six of these ships would see action in the Great War, although such was the rapid rate of technological development in this period that they were already virtually obsolete by 1914; three would be lost in action – HMS *Good Hope* and HMS *Monmouth*, both lost with all hands at the Battle of Coronel in November 1914, and HMS *Drake*, torpedoed by a German submarine off County Antrim in October 1917 with the loss of 18 lives.

It would be tempting to assume that a reluctant Royal Navy did not take its responsibilities seriously, but this is contradicted by the calibre not just of the warships provided for the manoeuvres but of their commanding officers, who included the brightest and the best of the Royal Navy's younger generation. Wilmot Fawkes was an old friend of Fisher, and a member of his inner circle, and two of his captains were Fisher protégés and rising stars of the Royal Navy. Charles Madden, who would serve as John Jellicoe's Chief of Staff at Jutland, commanded HMS *Good Hope* while Jellicoe himself was the Captain of HMS *Drake,* and would therefore have seen for himself some of the problems inherent in a large-scale enterprise requiring inter-service co-operation; this experience may have helped to play a part in his reservations concerning amphibious operations, such as those he expressed in a note to Winston Churchill on 8 January 1915 in relation to the First Lord's proposed assault on the North Sea island of Borkum.[22]

By 2:00 a.m. on Wednesday 7 September, the cruisers were off the coast at Clacton; the leading cruiser approached the shore at 5:30 a.m., with the ten transport ships about two miles from the shore extending in a line from opposite the pier for about three miles towards Frinton. In an interesting example, by no means the last, of the use of cutting-edge technology in British manoeuvres of this period, the senior Blue

21 The remaining transport ships were *Cestrian, British Prince, Drayton Grange, Antilan, Toronto* and *Consuelo.*

22 R. Prior, *Gallipoli, The End of the Myth*, p.43.

Embarkation of the invading force at Southampton, 1904 manoeuvres. (*The Graphic*, 19 September 1904, author's collection)

HMS *Drake*, part of the cruiser squadron in 1904. (Wikipedia)

umpire, Lord Methuen, was taken out to the flagship, *Good Hope*, on a speedboat, *Napier Minor*, one of a number provided to the War Office for the manoeuvres by the enterprising businessman Selwyn Edge. This gifted self-publicist was an Australian entrepreneur who worked with Napier & Son, making engines, automobiles and speedboats. He had become the first British winner of an international motor race in 1902 and the following year his Napier 1 speedboat won the inaugural British International Harmsworth Trophy in Cork Harbour. He would go on to become controller of the agricultural machinery department of the Ministry of Munitions during the First World War.

Half an hour after the transports had anchored, marines and engineers headed ashore in order to construct pontoons and landing stages. Steam pinnaces towed rafts, poles and timbers to construct temporary landing gangways. At around 8:00 a.m. the first troops began to come ashore in landing craft, men of the 2nd Wiltshires and 1st Yorkshire Light Infantry disembarking from the *Marquette*. One journalist remarked that 'the troops looked very fit and capable, and their thirty-six hours at sea, cooped up on the transports, had done them no visible harm.' Many of the horses did not fare so well – two or three had to be shot and thrown overboard.[23]

Disembarkation went on all day, 2nd Division landing on the beach at Clacton and 1st Division along the coast at Holland Gap. By the afternoon the early morning mist had been replaced by bright sunshine. By this time the tide had gone out, presenting problems as fully laden boats could get no nearer than 100 yards from dry land, and the decision to land the infantry first meant that the mounted troops, guns and baggage had to be landed in knee- or waist-deep water. Many horses and their riders fell into the water, causing 'roars of laughter among the thousands of spectators.' The local press provided much mirth through its portrayal of a full-scale landing on a beach full of holidaying Londoners: 'The landing craft went on, while bathers in costumes of marvellous tints bobbing up and down in the water made a comical picture amid the landing stages and khaki-clad troops.'[24] At all times huge crowds watched the events, with civilian and military police preventing sightseers from getting too close to the action. Special trips from London by the Belle steamers were chartered for those wishing to view the fleet. Fish stall proprietors did a roaring trade from troops and spectators alike. Some soldiers slept on the beach while others entertained themselves with impromptu concerts.

Disembarking continued throughout the night, relying on searchlights from the cruisers and high power electric lights and flares from the land. By the evening the weather had taken a turn for the worse, with heavy rain showers and a fresh south-westerly wind making things difficult for the soldiers and sailors unloading the pontoons. By the next morning many of the boats had been completely smashed and others damaged, a powerful reminder of the threat from the weather to such an operation. Casualties were remarkably few given the numbers involved and the

23 *The East Essex Advertiser and Clacton News,* 10 September 1904.
24 Ibid.

Sailors from HMS *Kent* assisting the landing of men of the 2nd Wiltshire Regiment. (*Illustrated Budget*, 17 September 1904, author's collection)

The 1st Dragoon Guards landing their horses. (*Illustrated Budget*, 17 September 1904, author's collection)

Spectators watching the men of HMS *Kent* floating landing stages. (*Illustrated Budget*, 17 September 1904, author's collection)

Landing at Clacton. (*The Graphic*, 17 September 1904, author's collection)

complexity of the operation. One sailor was knocked out by a pole and taken to the cottage hospital ashore. Another sailor was kicked by a horse and severely injured. A crew man on HMS *Good Hope* was 'dangerously injured by a boom which struck him in the back', with little hope for his recovery.

The landing had not proceeded entirely smoothly. The fact that opposition to the landing might occur was ignored. No attempt had been made to secure the landing places, for example by sending small parties of men in the first boats. Units did not land in places opposite the positions where their transports had anchored, causing longer journeys than necessary. A lack of staff officers onshore to direct the men meant that the infantry landing took longer to move away from the shore than they should have done. Some units were landed in the morning without their first line transport which didn't reach them until evening. Lyttelton commented acidly in his report that 'It is difficult to understand this procedure except on the assumption that it had been allowed to transpire throughout the invading force that the landing would be unopposed.' He had decided against allowing the Reds to interfere with the landing 'because in the face of such opposition a landing would probably have been quite impossible' (which rather called the point of the whole exercise into question) but argued that such interference should not have been ignored, it should have been expected.[25] This was another example of the confused thinking at the heart of the 1904 manoeuvres.

25 TNA WO 279/8, Report on Army Manoeuvres 1904, p.24.

Landing guns on the beach at Clacton. (*The Graphic*, 17 September 1904, author's collection)

While disembarkation proceeded, the first troops ashore formed up on the greensward above the cliffs and headed inland. The first clash between the two armies came when the Royal Scots Fusiliers rushed the village of Little Clacton, nearly cutting off a defending force from the 5th Lancers. The defenders were surprised when the Fusiliers advanced under cover of thick hedges, the first sign of a recurrent feature of these manoeuvres, the effect on combat of the enclosed country in which they took place. One officer narrowly escaped capture when three Fusiliers sprang on him out of a hedge. The Lancers retreated along the Colchester road, halting every 200 yards to fire on the enemy.

That night, French's leading units attacked Red's 6th Brigade in a confused night action around Great Bentley. This action was later criticised by the Red umpire, General Olyphant, as not according with the stipulation that manoeuvres should be as realistic as possible. An invader would not have attempted such a course shortly after landing in a strange country, much of it enclosed, with a hostile population and a strongly entrenched enemy which knew the ground and roads well. He also felt that it was unlikely that the defeated 6th Brigade would have been forced to retire seven miles, as they were ordered to do by the umpires, thereby abandoning Colchester.[26] This is a good example of the impact a contentious umpiring decision could have on the course of manoeuvres.

26 Ibid., p.83.

Early on the next morning (Thursday 8 September) the invading Blue force marched into Colchester. At this point the invaders were worn out, and to add to their discomfort heavy rain began to fall. An armistice was – to the evident relief of all – announced, and the exhausted troops, seeking such shelter as was available, were soon asleep. While understandable, this pause added a further note of unreality into proceedings, especially when the invaders were up against the clock, as would soon become apparent. Operations resumed on Friday 9 September, French advancing south-west from Colchester as far as Witham, while Wynne, reinforced by the arrival of the Guards' London Brigade, was playing a waiting game at Braintree, holding the road to London.

It was at this point that the directing staff intervened to change the terms of the operation in a good example of the highly circumscribed approach to manoeuvres favoured at the time. A telegram was delivered to Sir John French, informing him that the Blue invading force which had landed in Sussex had been defeated and driven into the sea, and ordering him in the circumstances to fall back to his transports and put to sea as rapidly as possible. The retirement was begun at once, the troops heading back through Colchester. During the retreat the Blue Cavalry Corps at Ardleigh suffered heavy losses at the hands of Colonel Allenby, but with that exception French's force managed to retire without loss. Sunday was spent quietly on both sides, with the Blue troops in their camps at Little Bentley and Alresford while Wynne was at Colchester. It was at this point that the sole recorded fatality from the manoeuvres occurred, when Driver Francis Harris of the Army Service Corps was killed when driving a water cart to a tank. A trace broke, the horses bolted and the cart turned completely over and fell on him, the wheels passing over his head. The jury at an inquest in Colchester the next day returned a verdict of 'accidental death' and the 19 year old Harris was buried with military honours in Colchester Cemetery.[27]

On Monday Wynne's pursuing force caught up with the retreating Blue troops at Weeley Heath, and a hotly contested battle ensued. Sir John French succeeded in entrenching himself in positions stretching from Brightlingsea to Beaumont, thus protecting the whole of the Naze from assault. *The East Essex Advertiser* observed that it was impossible to see more than 300-400 yards at any time, usually much less, describing an 'extremely instructive (fight) in a country that is probably as densely cultivated as any part of Great Britain'.[28] In these conditions cyclists were thought to be more useful than cavalry. After brisk fighting around Thorpe le Soken, French withdrew, leaving General Bruce Hamilton to cover the embarkation. After a 'long-continued and magnificent struggle' General Wynne's attack was adjudged to have been repulsed with heavy loss all along the line.[29] *The East Essex Advertiser* was critical of the umpires' rulings as Wynne had inflicted heavy losses on Hamilton, with

27 *The Essex County Chronicle*, 16 September 1904.
28 *The East Essex Advertiser and Clacton News*, 17 September 1904.
29 *The Essex County Chronicle*, 16 September 1904.

Allenby's cavalry prominent in the attack: 'It is somewhat difficult to understand how it could be asserted that he had been driven off, for Col. Allenby had undoubtedly succeeded in breaking through the defence on the left flank, and as well in the rear of the position, and threatening the line of communication with Clacton.'[30] It is worth noting, however, especially in light of his performance in later manoeuvres, that Allenby's pursuit of the retreating Blue force was criticised by the Chief Umpire, Connaught, in his report: 'The role of the Red force was to strike at the enemy wherever found with the utmost energy. From the general conduct of the pursuit this principle does not seem to have been sufficiently borne in mind.'[31]

Following a delay imposed by tides and adverse weather conditions, the Blue force began to re-embark on Tuesday 13 September in conditions of bright sunshine and a calm sea. Once again there was a deep throng of spectators to observe the process, which went more smoothly than disembarkation a week before. For example, 93 Battery, Royal Field Artillery was embarked rapidly and the entire battery put afloat in the space of twenty minutes. Troopers embarked with their horses, 10 per boat, for the most part very calmly. The pontoons were towed out to *Atlantian* and hoisted aboard with the assistance of the Royal Engineers. The wind got up a little in the afternoon and there were a few showers of rain. One horseboat sank, though with no losses. The embarkation ended at midnight on Wednesday, when the 8th Hussars were the last troops afloat, leaving the 1st Dragoon Guards to set off from West Clacton camp on Thursday for Aldershot by road.

By Thursday morning the fleet had departed and the beaches were once more in the hands of the holidaymakers. The transports and their cruiser escorts reached Southampton on Friday morning. For the most part, Clacton was sorry to see the end of the manoeuvres – one local paper lamented that 'Never again shall we see the Duke of Connaught and his General Staff and a brilliant group of military attachés, naval and military, standing on Clacton beach amidst a seething crowd of East End trippers and mountebanks, and nigger minstrels, and shell-fish vendors.'[32] One exception was the bathing machine operators, who complained about the loss of business, as bathing was entirely suspended and their machines requisitioned by officers. Other than that, the only complaint in the local press was that 'owing to the strong counter-attractions on the beach, the attendance at the Football Club benefit concert, kindly given by Mr Forsyth on Wednesday, was not so large as could be wished.'[33]

The manoeuvres were attended by a large number of foreign military attachés, who stayed in the George Hotel at Colchester and were transported daily in a fleet of privately owned cars specially lent for the occasion. They included the U.S. attaché, Major J.H. Beacon, the Russian Major-General Yermoloff, who 'in his huge car, with

30 *The East Essex Advertiser and Clacton News*, 17 September 1904.
31 TNA WO 279/8, Report on Army Manoeuvres 1904, p.74.
32 Kenneth Walker, *The History of Clacton*, p.43.
33 *The East Essex Advertiser and Clacton News*, 10 September 1904.

Plaque at Holland-on-Sea marking the landings. (Author's collection)

the French military attaché, was a hazard to all and sundry as he appeared and disappeared in a cloud of dust' and Major Count von der Schulenberg, later Chief of Staff to Crown Prince Wilhelm.[34]

The most historically important of the attachés was the French representative Colonel Albert d'Amade. He is probably best remembered, at least by the British, for commanding the three French territorial divisions to the left of the BEF in August 1914. More significantly in light of his experiences in 1904, he went on to command the French contingent, the *Corps expéditionnaire d'Orient*, during the initial stages of the Gallipoli Campaign. This appointment was not the accolade it might at first sight appear; General d'Amade had been relieved of command on 17 September 1914 and was generally held to have performed poorly, his three divisions being 'lost' after they retired from Amiens. His appointment to the command of the CEO indicates the low priority Marshal Joffre gave to the Dardanelles expedition as much as his desire to get an unsatisfactory commander out of France.[35]

In the event d'Amade was adjudged a failure in the Dardanelles, the British commander General Sir Ian Hamilton finding him to be defeatist and pessimistic,

34 T.A Baker, *Clacton-on-Sea in old picture postcards volume 1* p.44.

35 E. Greenhalgh, *The French Army and the First World War* p.102. For d'Amade's career see Colonel Emile Mayer, *Nos Chefs de 1914* (Paris, 1930). Mayer commanded the artillery in one of d'Amade's territorial divisions in 1914.

and he was replaced by General Henri Gouraud, at 48 the youngest general in the French army, on 15 May 1915.[36] For his part, d'Amade was extremely critical of the British method of operations at Gallipoli, in particular War Office interference with the commander in chief on the spot, and it is possible to speculate that the roots of his disillusionment with the British way of waging war may have been sown on the beaches of East Essex eleven years before.

The Secretary of State for War and Commander-in-Chief could be satisfied with many aspects of the 1904 manoeuvres. They had carried out an amphibious landing and re-embarkation with a sizeable body of troops and had secured naval participation in the operation, even if the Admiralty's co-operation had been grudging at best. They had also tested the army's ability to conduct manoeuvres away from Salisbury Plain – as Arnold-Forster declared to the Prime Minister, 'We have had manoeuvres – at last – in enclosed country, which really represents five-sixths of England.'[37] French expressed his reservations in a letter he wrote to Lord Esher on 13 September, lamenting the fact that the scheme had not been carried out as originally planned – if it had, 'It would have been cheaper in the end, for it would have been a more valuable experience, and would probably have satisfied the public, which these very half-and-half measures have apparently quite failed to do.'[38]

To modern readers, interest in the 1904 manoeuvres lies mainly in the fact that they featured an amphibious landing on open beaches which appears to be a precursor, in however rudimentary and limited a form, of the Gallipoli landings eleven years later, but this element excited little public comment at the time. Instead, contemporary commentators were keen to discuss the terrain, which presented such a contrast to previous manoeuvres on the open spaces of Aldershot and Salisbury Plain. At a meeting of the RUSI on 31 January 1905, with Colonel Henry Rawlinson, Commandant of the Staff College, in the chair, Colonel G.H. Ovens argued that there was an urgent need to discuss the implications of fighting in enclosed country such as had recently been encountered in Essex.[39] Ovens was well placed to comment, having commanded an infantry battalion in the 1903 manoeuvres and acted as Chief Staff Officer of the Red Umpires in 1904.[40]

This RUSI paper has been accorded little attention by historians, but it is highly significant as the most detailed analysis of any set of army manoeuvres of the period, even if its failure to address the topic of amphibious operations may now seem perverse. Ovens was far more concerned with the inferences that could be drawn for

36 G. Cassar, *The French and the Dardanelles: A study of failure in the conduct of war*, pp.122-123.
37 Moon, op. cit., p.254.
38 French, op. cit., p.152.
39 Colonel G.H. Ovens, 'Fighting in Enclosed Country with Some Notes from the Essex Manœuvres', *Royal United Services Institution Journal*, pp.524-546.
40 Ovens had commanded 1st Battalion, the Border Regiment in South Africa and was later briefly (September-November 1914) in command of 68th Infantry Brigade.

fighting in enclosed country, whether on the continent or closer to home – Britain had to abandon 'the ostrich-like plan of shutting our eyes and ears to the possibility of an invasion of England. Great events move with accelerating rapidity in this twentieth-century, and unexpected developments are the rule.' [41]

Ovens noted that in flat countryside, with small fields and thick hedges which were in full leaf in September, it was impossible to see any distance, or to know what was going on even 150 yards away. East Essex was 'a land of surprises and ambushes'.[42] One commanding officer moving with his battalion had told Ovens that he suddenly couldn't find his men, and took several minutes to find the battalion again, even though it was never more than 200 yards away from him in another field. Heads of columns, patrols and orderlies were constantly losing their way. In such circumstances, roads became all-important, but this inevitably meant congestion since cavalry, guns, ammunition columns and transport were all confined to them, and all but small infantry units were dependent on them. Ovens therefore suggested that infantry scouting was the most dependable type of reconnaissance, along with the use of cyclists, and he further recommended personal reconnaissance by commanders. Given that 'on the level they cannot see anything', he suggested that 'Generals commanding may climb church towers, haystacks, and trees, or even the new self-raising fire escape ladders.'[43] It is interesting to speculate whether this is the advice General Morland had in mind on 1 July 1916 when he took up position in a tree in order to follow the assault on Thiepval, an action for which he has been much derided. Morland was not present to hear Ovens' paper, though his army commander at the Somme, Rawlinson, was.[44]

Ovens' conclusions appear in many ways to have been remarkably prescient when it came to the trench warfare of the First World War, which was in effect an alternative means of negating the possibility of mobile warfare. He stressed the importance of defensive works, recommending that the Royal Engineers should weave barbed wire and wire charged with electricity into hedges and gateways and block off roads with abattis, while banks could be mined and gaps cut to create gateways on which machine guns would be trained. Hand grenades should be issued and shields provided for men holding important exposed positions. The army should also consider such innovations as automatic rifles, hand howitzers and bamboo mortars.[45] In attempting to prescribe the most effective tactics for defence in enclosed country, he seems to have unconsciously described trench warfare methods from the coming war.

41 Ibid., p.525.
42 Ibid., p.528.
43 Ibid., p.529.
44 As well as Rawlinson, those who contributed to the discussion following Ovens' paper included Major-General Edward Hutton, Colonel Edward Altham and Lieutenant Colonel A.W.A. Pollock.
45 This was a weapon used by the Japanese in Manchuria, which did not catch on in other theatres of war.

Arnold-Forster felt that the manoeuvres had conclusively proved the impossibility of an invasion of Britain, reporting to Lord Kitchener in the following terms:

> I saw the embarkation, which occupied four days, interrupted by the occurrence of a summer swell, I saw the Men-of-War and transports lying at anchor by day and by night within one hour's steaming of the Harwich Flotilla of destroyers on one side, and the Medway Flotilla on the other, and I knew well that if it had been real war, not one of the ships that lay off Clacton would have been afloat an hour after sundown. But, indeed, I did not need the object lesson of the Essex Manoeuvres to convince me that our present Military organisation is a folly; that as long as our Fleet swims there will be no invasion, and that when our Fleet fails us, no force on land, however numerous, however well organised, will preserve us from ruin.[46]

According to the manoeuvre report, the Army had learnt that a force of 10,000 men and 2,000 horses of all arms could be landed in ten hours and that within twenty-four hours, sufficient transport could be landed to keep this force in the field for two or three days (though, of course, this had been an unopposed landing). They also learnt a number of detailed lessons for the future, such as the need for a military staff officer (with a staff of cyclists and signallers) to be assigned to each naval beach master.[47] Above all, the lesson of the 1904 manoeuvres was that 'It is very questionable whether, in these days of improved firearms, a descent on a coast line can be carried out in the teeth of a resolute opposition.'[48]

The press was distinctly unimpressed by the manoeuvres, and specifically their lack of realism. The journalist and future crime writer Edgar Wallace, writing in *The Daily Mail,* dubbed Clacton 'the Capital of Letspretendia'.[49] T. Miller Maguire, a strong advocate of home defence, was even more derisive, describing the proceedings as:

> A few peregrinations in the East of England, devised by the dialecticians of the Ministry… planned, not to train the soldiers but… to play upon a credulous public – to give the people a show for their money – they want to know how their money is spent; well, most of it goes in nonsense like this. A HYPOTHETICAL INVASION. Invasion of what? England or Laputa or Timbuctoo? No power on earth will ever practice (sic) invasions under the Essex conditions – hence the Midsummer Night's comedy of the whole transaction.[50]

46 Moon, op. cit., p.257.
47 TNA WO 279/8, Report on Army Manoeuvres 1904, p.25.
48 TNA WO 279/8, Report on Army Manoeuvres 1904, p.17.
49 Norman Jacobs, *Clacton Past with Holland-on-Sea & Jaywick*, p.47.
50 Moon, op. cit., p.259.

Maguire and Strachey were disheartened to learn that one of the main lessons of the manoeuvres was that the problems of fighting in enclosed country would be as serious for the defenders as for the attackers, thereby casting doubt on the ability of a 'hedgerow defence' to resist an invasion – Charles à Court Repington, the Military Correspondent of *The Times*, was dismissive of 'the stock theory that a few half-trained men in the country hedgerows can hold a large force in check for a large period'.[51] He was also critical of the decision to stage an unopposed landing with no element of surprise, writing to a War Office friend that 'I shall most likely look in at the landing on the coast. I presume the place and time of the enemy's attack will be duly noted in the orders of the defending army!'[52] *The Graphic* adopted a more positive note – 'The army manoeuvres in Essex, though in some measure disappointing, have been, on the whole, instructive' – though conceding that 'There was, perhaps, a larger amount of "make believe" in the scheme than there is sometimes in such operations.'[53]

The Army and Navy made serious and concerted efforts to learn from the experience of the Clacton landings. A joint conference held on 20 March 1905 discussed the details such as transportation, embarkation and disembarkation, and the staff work and logistics required for large-scale combined operations. Doubts were expressed over the opposed landing on an open beach of troops under the conditions now prevailing, since 'modern weapons had greatly added to the difficulties of a force attempting an opposed landing.' The conclusion of the meeting was that 'enough has been said… to demonstrate the impracticability of landing troops now-a-days in the face of opposition.'[54]

This assumption was enshrined into British doctrine in the War Office's *Manual of Combined Operations* (1913), which provided the main source of information and guidance for a commander finding himself planning an amphibious operation, as Sir Ian Hamilton did in 1915.[55] Hamilton, who did not participate in the 1904 manoeuvres as he was observing the war in Manchuria, took with him to the Dardanelles a copy of the 1913 Manual; Orlo Williams, Hamilton's cypher officer, confirms that it was studied by all the officers on Hamilton's staff as well as the officers of the formations under his command, and its principles were followed as far as possible.[56] However, Hamilton's chief engineer was critical of the procedures set out in the Manual in his *Notes on Landing*, written on 10 May, after the initial landings at Gallipoli. The Manual was far stronger on the minutiae of beach masters and landing parties than the tactics required for amphibious warfare, and Edward Erickson points out that

51 *The Times*, 12 September 1904.
52 Moon, op. cit., p.253.
53 *The Graphic*, 17 September 1904.
54 D. Morgan-Owen, Lessons Learned from Gallipoli, http://defenceindepth.co/2015/05/11/lessons-learned-from-gallipoli (accessed 30/8/2015).
55 Ibid.
56 P. Chasseaud and P. Doyle, *Grasping Gallipoli: Terrain, Maps and Failure at the Dardanelles, 1915*, p.103.

this was not a uniquely British failing; no nation at the time had a formal doctrine of amphibious warfare.[57]

The Manual spelled out clearly that the key to success for amphibious operations was, unsurprisingly, co-operation between the two arms:

> The complicated duties of embarking and landing troops and stores can only be carried out successfully so long as perfect harmony and co-operation exist between the naval and military authorities and commanders, and when the staff duties devolving on both services have been carefully organised and adjusted.[58]

Following the Clacton exercise, a few key individuals made some attempt to create 'perfect harmony and co-operation' between the two arms, so that it would be wrong to think that the 1904 manoeuvres killed off any trace of inter-service co-operation on combined operations. George Aston, Deputy Assistant Adjutant General at the Staff College 1904-1907, began a dialogue with his naval counterpart, Edmond Slade (Commander of the Naval War College at Greenwich from 1904) which bore fruit in student exchanges, joint lectures and combined staff tours of coastal areas. These first steps were supported by the Director of Naval Intelligence, Battenberg, who argued that 'The more we can instil sound ideas of Naval War and the *practical* possibilities of joint action between the two Services into the minds of rising military men, the better.'

The joint staff tours, which started in 1905, involved the various stages of an amphibious operation, a night-time landing by an advanced guard, a full disembarkation at dawn and the establishment of a secure beachhead. Valuable experience in planning combined operations at a staff college level was gained, for example in the staff tour held in April-May 1907, which included practising an opposed re-embarkation from Sandown Beach involving Royal Marines and half a battalion of the Royal Fusiliers.[59] Following the tour, Henry Wilson, the Commandant of the Staff College, expressed the hope that future tours could feature the disembarkation of men, horses, guns and vehicles in order to enhance the practical value of the exercise. However, these tours ended the following year thanks to the move to a continental commitment for the British Army which now held sway in the highest circles, even though Slade had insisted that further joint training was essential given that 'we cannot move a single man to the theatre of war, wherever it may be, without carrying him across the sea.'[60] Aston continued to press his views, for example in a lecture entitled 'Amphibious

57 E. Erickson, *Gallipoli: Command under Fire*, p.111.
58 TNA WO 33/611, Manual of Combined Naval and Military Operations September 1913, p.293.
59 TNA WO 33/2982, Joint Report on Naval and Military Staff Tour 1907.
60 Ibid., p.4.

Strategy' to the RUSI in July 1907, but his chance to influence national defence policy had passed.[61]

Another example of a determined individual attempting to forge inter-service links was the Admiralty's Assistant Director of Naval Intelligence, Commander George Ballard, who had taken part in the joint Conference on Overseas Operations following the Clacton manoeuvres. He opened a correspondence in August 1905 with Charles Callwell, Assistant Director of the General Staff's Directorate of Military Operations, on the feasibility of combined Baltic operations, specifically an expedition to Schleswig-Holstein, though Callwell had to inform Ballard in October 1905 that after detailed study of the practicalities of such an operation, the army had abandoned support for a strategy that one officer had derided as 'Fisher's invasion of Germany' in favour of plans for sending an expeditionary force to the Franco-Belgian border in support of the French. Callwell, like Wilson and Grierson, much preferred a strategy which gave the army an independent role to one which cast it as an instrument of the navy.[62]

One illustration of the difficulty of getting the Army and Navy to work together comes from Edmonds' time as Chief of Staff to Major-General Thomas Snow, commander of 4th Division between 1911 and 1914. Snow was keen to give his troops practice in embarking and disembarking at a pier and on a beach, and so hired an old cross-channel paddle steamer which was to transport 11th Brigade from Colchester, landing it on the Isle of Thanet, where it would fight 10th Brigade. The journey was made, the sea was nearly dead calm and Snow and his staff were waiting on the beach for the landing to proceed, only for the naval representative to insist that, as the glass was falling, it was not safe to attempt landing troops from boats. So although the Brigade's commander, Henry Heath, was willing to take the risk, Snow reluctantly had to agree to the troops being taken around to Folkestone and landed at a pier instead to skirmish with the Shorncliffe and Dover brigades.[63]

What links can be drawn between the landings between Clacton and Holland-on-Sea in September 1904 and the amphibious operation at Gallipoli eleven years later? The idea of attacking Turkey via the Dardanelles Straits had been current since the late nineteenth century, and a month after the 1904 manoeuvres, French submitted a report on the area's strategic significance (*Report on the Bosphorus and Dardanelles, October 1904*) which was well received by Fisher at the Admiralty, but little was done in the way of planning the 'Forcing of the Dardanelles' before January 1915, when a naval operation was envisaged rather than joint action between Army and Navy.[64]

The most noteworthy link between the two operations in personnel terms is Sir Frederick Stopford, intimately involved in the planning for the 1904 manoeuvres as

61 S. Grimes, *Strategy and War Planning in the British Navy, 1887-1918,* p.68.
62 Ibid., pp.68-70.
63 Brigadier-General Sir James Edmonds, *Memoirs,* chapter XXII p.6.
64 P. Doyle, *Gallipoli 1915,* p.21

Director of Military Training and a key figure in 1915 as the commander at Suvla Bay. In the view of Edmonds, Stopford had 'failed miserably' when tested at manoeuvres and would never have been given a command in peace time.[65] In an article written in 1930, when he felt able to speak more freely than in the Official History, Edmonds argued that the failure to find adequate generals for the Gallipoli campaign was due to 'a system which refused to maintain a sufficiently large army and to attract brains to it'.[66] Edmonds may have had in mind Stopford's humiliation during the 1907 manoeuvres, when as Red commander he was captured, along with Brigadier-General Samuel Lomax, by men from the 8th Hussars, and had to be released so that he could continue to deploy his troops.[67] It is interesting to note that Stopford's opponent in 1907 was none other than Sir Ian Hamilton, who would be his commanding officer at Gallipoli. Certainly his experiences in manoeuvres do not appear to have equipped Stopford in any way for his role in August 1915, and it is perplexing that a man with his total lack of experience in leading a sizeable body of men in combat should have been appointed to command at Suvla Bay at all.

A number of other individuals who played a significant part in the Gallipoli campaign gained experience of a large-scale amphibious operation at Clacton. One was Captain Walter Braithwaite, who acted as a Red Assistant Umpire, and would later serve as Sir Ian Hamilton's Chief of Staff at Gallipoli. Another was Colonel Edward Altham, who was on the Blue umpiring staff in 1904 and took a leading role in the discussion which followed Ovens' RUSI paper.[68] He was already seen as one of the foremost military thinkers in the army, and would later be responsible for the influential *Principles of War*, probably the clearest statement of the strategic thinking of the British General Staff on the eve of war, for which his friend Horace Smith-Dorrien wrote the foreword.[69] Altham, a former Head of the Strategical Section at the War Office, ultimately rose to become a Lieutenant-General and was sent out to Mudros in July 1915 to resolve the administrative chaos following the initial landings at Gallipoli. In his role as Inspector-General, Communications to the expeditionary force Altham encountered problems in dealing with the Navy reminiscent of 1904: 'Our different organisations and systems want fitting into each other and at times the Navy system does not seem to us altogether well-adapted to the particular job we are both at now.'[70]

The 1904 exercise was a worthwhile and imaginative scheme, but one which was sadly bungled both in planning and execution, something which could also be said for

65 A. Green, *Writing the Great War – Sir James Edmonds and the Official Histories 1915-1948*, p.53.
66 J. Edmonds, 'Old Men at Suvla', *The Army and Navy Gazette*, 8 November 1930.
67 T. Crawford, *Wiltshire and the Great War: Training the Empire's Soldiers*, p.18.
68 Ovens, op. cit., pp.541 543.
69 M. Howard, 'Men Against Fire: The Doctrine of the Offensive in 1914' in *Makers of Modern Strategy from Machiavelli to the Nuclear Age* edited by Peter Paret, pp. 518-519, 909.
70 R. Crawley, *Climax at Gallipoli: The Failure of the August Offensive*, p.147.

The beach at Holland-on-Sea, where 1st Division landed. (Author's collection)

the Dardanelles eleven years later. At this stage, two years after the end of the Boer War, the British army was still struggling to come to terms with large-scale manoeuvres. Many complex tasks were beyond it, not least an ambitious amphibious landing requiring navy-army co-operation. Once the decision had been taken not to provide for an opposed landing, much of the point of the exercise was lost. Experience in disembarking and embarking a sizeable force of men was gained, though as *The Times* pointed out, it was a matter of concern that even in an uncontested landing carried out in favourable weather, boats had been wrecked and stores lost.[71] The 1904 manoeuvres represent a missed opportunity to carry out meaningful preparation for an amphibious operation, but the fundamental causes of the Allied failure at Gallipoli were less to do with pre-war planning and more to do with the hasty preparations behind an army operation which was forced on it by the failure of the naval attack of 18 March 1915, along with a fatal underestimation of the strength of the defences and the fighting qualities of the Turks. It is a matter of speculation whether any long-serving men of the 2nd Royal Fusiliers who landed on X Beach on 25 April 1915 were reminded as they made their way ashore of a day in September eleven years before, when they had been among the first troops ashore at Clacton.

71 *The Times,* 13 September 1904

3

# Manoeuvres 1904-1912

After the experimentation of the 1904 manoeuvres, things returned to normal with the 1905 manoeuvres taking place on Salisbury Plain, a pattern repeated in future years. Each year the programme proceeded in incremental stages. For example, the 1908 manoeuvres began with a Staff Tour, held in the West of England between the River Severn and Reading between 8 and 12 September. This imagined a Blue Western invading army (three brigades of mounted troops and three divisions) landing at Avonmouth and Portishead, against a Red defending army. The objective was 'to afford officers practice in dealing with higher organisations than those with which they are usually accustomed to deal'.[1] This was followed by a two-stage set of army manoeuvres, an indication of the British Army's growing ambitions and sophistication. In the first phase, which took place from 14 to 16 September, the 1st and 2nd Divisions of the Aldershot Command were opposed to each other, each with a brigade of mounted troops. The Blue Army, commanded by Major-General James Grierson, was stated in the 'General Idea' to have landed at Southampton and was opposed by Major-General T.E. Stephenson. Smith-Dorrien acted as Chief Umpire, assisted by William Robertson. For the second phase of operations (17 to 19 September) Smith-Dorrien commanded Red Force with Grierson and Stephenson each commanding a division under him, opposed by a small force of 2,500 Blue troops under Colonel Julian Byng.

Grierson was adjudged to have done well in the 1908 manoeuvres, out-marching the Red forces to a key ridge and forcing Stephenson to spend the rest of the second day in piecemeal attacks trying to dislodge him. Less successful was a young and ambitious cavalry officer, Colonel Hubert Gough, whose entire cavalry brigade was taken by surprise on the first night of operations when fast asleep and captured by the 2nd South Lancashires. The umpire on the scene contacted the Chief Umpire,

1 TNA WO 279/21, Aldershot Command Staff Tour and Manoeuvres, 1908.

Smith-Dorrien, who said that it was quite clear – the cavalry brigade was considered to be out of action.[2]

1909 was on a significantly greater scale than the previous year, with Smith-Dorrien's Aldershot command pitched against a Blue force under Sir Arthur Paget made up of the Eastern and Southern Commands. Smith-Dorrien, commanding two divisions led by Grierson and Stephenson as well as a half division of cavalry under Allenby, could call upon 22,685 men, slightly more than Paget's Blue army, which consisted of two divisions under Franklyn and Belfield and a small cavalry force under Fanshawe.[3] The manoeuvres took place in the Cotswolds between Cheltenham, Swindon and Oxford, and were supposedly based on the Novara campaign of 1849, in which the Austrian Field Marshal Radetzky had defeated the army of Piedmont. According to the War Office report, which included a map and narrative of the Novara campaign, the Thames substituted for the river Ticino in the original, Swindon for Mortara, Cheltenham for Milan and Oxford for Pavia.

It could be asked what relevance a campaign fought with muskets and muzzle-loading cannon from sixty years before had to the new conditions of warfare; it is hard to avoid the suspicion that the inspiration for the scheme, drawn up by the Director of these manoeuvres, Sir John French, had a great deal to do with the chapter on Novara in Hamley's *Operations of War*, the sole text used in the Staff College entrance examination until 1894 and, according to Edmonds, the only book French ever borrowed from the War Office Library, though Edmonds added with characteristic cattiness that French had failed to understand it. French's plan was that the first day (21 September) would be used to concentrate the two armies, the second day for reconnaissance and the last day for the decisive battle. Unfortunately he did not take into account the marching ability of the troops, who covered twenty-three to thirty-five miles instead of the expected fifteen, or the aggressive approach of both commanders, with the result that by the end of the first day his schedule was already irrelevant.

French had encountered a problem which also dogged armies on the continent. One danger inherent in the limited duration of manoeuvres was that commanders keen to make an impression and to advance their careers were prepared to exhaust their troops in the knowledge they could rest when the manoeuvres ended, a luxury not open to them in real combat. One Austrian division carried out a forced march of 40 miles in 24 hours, three times the usual rate of march, and Austro-Hungarian manoeuvres regularly resulted in the deaths of troops, many from heat stroke.[4]

This was the first set of army manoeuvres under a new umpiring system, and in particular the introduction of screens to represent casualties, and complaints were made that although they represented an improvement, many troops took time to

2 R.R. Seim, *Forging the Rapier among Scythes: Lieutenant-General Sir Horace Smith-Dorrien and the Aldershot Command 1907-1912*, p.95.
3 TNA WO 279/31, Report on Army Manoeuvres 1909.
4 L. Sondhaus, *Franz Conrad von Hötzendorf: Architect of the Apocalypse*, p.92.

Images from the 1909 manoeuvres near Stow on the Wold. (*The Illustrated Sporting and Dramatic News*, 25 September 1909, author's collection)

adjust to them; some, for example, were ready to unfurl their screens when ordered to do so by the umpires but then forgot to keep them clearly displayed, especially after the unit had moved. The official report recommended that the screens should be bigger and so more visible, and should have metal rather than wooden supports so they could be driven into hard ground; there was also a problem with many of the wooden supports ending up on the camp fire in the evening. Furthermore, it was suggested that in future there should be screens of one colour to represent the effect of infantry fire and another for artillery fire. However, this proposal was not acted upon; in the first place it would require more screens to be carried, and it was also disputed whether troops needed to know whether fire superiority was being gained more by infantry than artillery.[5]

Some umpires expressed reservations about ordering a body of troops which had been careful to conceal itself to put up a screen, on the grounds that this could give away their position. In general, however, it was agreed that once the troops had become accustomed to the screens, 'they will be a very valuable means of improving the training. They have a marked influence on the speed of operations, and they supply practical means of demonstrating the effect of fire to local commanders.'[6]

The Blue Senior Umpire, Douglas, was severely critical of the staff work on display and of 'a tendency on both sides to rush hastily into battle', adding that 'There was

5 TNA WO 279/31, Report on Army Manoeuvres 1909, p.110.
6 Ibid.

A scene from the 1909 manoeuvres, showing the new casualty screens in use. A sketch by the Dutch artist H.W. Koekkoek. (*The Illustrated London News*, 25 September 1909, author's collection)

no cohesion in the various attacks or engagements… subordinate commanders were frequently unaware of what they were required to do, and rarely knew what other parts of the force were doing.'[7] Douglas, a somewhat conventional thinker described by Edmonds as 'a man with the talents of an orderly-room clerk' revealed his Staff College training in his insistence that insufficient stress had been placed on the various stages of a battle, and that they had not been kept sufficiently distinct: the preparation, the approach, the building up of a firing line at the final fire position and finally the decisive assault.[8]

The following year, Lieutenant-General Sir Herbert Plumer commanded the largest single force yet seen in British manoeuvres, which took place between Salisbury and Winchester between 21 and 23 September. His Red army, based on the Aldershot Command, consisted of 27,302 men, made up of one infantry division, a sizeable territorial detachment and two cavalry brigades, one regular and one territorial. Douglas' Blue force was smaller (20,662 men) and comprised two infantry divisions and a cavalry half division. In another instance of the unavoidable truism that in a

7 Ibid., p.112.

8 See J. Terraine, *Douglas Haig – The Educated Soldier*, pp.47-48 and G. Sheffield, *Douglas Haig: From the Somme to Victory*, pp.27-30 for the way in which Haig's views were similarly shaped by his own Staff College experience.

Winston Churchill and Sir John French at the 1910 manoeuvres on Salisbury Plain. (Soldiers of Oxfordshire Museum)

small army the same names were bound to crop up in different combinations each year, Smith-Dorrien (Red) and Paget (Blue) acted as the Chief Umpires, with French once more the Director.

Although there were the same old complaints about the poor quality of the entrenching by some troops, the umpires' reports suggest a better performance overall than in 1909; there was greater co-ordination between the different arms and subordinates were prepared to detach troops to assist their colleagues. The two commanders did not emerge from French's final report so well – Plumer was criticised for dispersing his forces too widely, while Douglas was faulted for launching piecemeal attacks. Interestingly, given the reputation he would later earn on the Western Front for thorough planning and methodical execution, Plumer was judged by Smith-Dorrien to have taken 'great and unnecessary risks' in attempting to carry out his plan of envelopment. Even so, it is possible to see 1910 as confirming that the British Army was at last getting to grips with manoeuvres, French's report noting that 'satisfactory indications were everywhere observed of the conception of an army as one organism.'[9]

The 1910 manoeuvres represented a further step forward, not just in terms of scale (a total of 48,000 troops engaged) but the first appearance at British Army manoeuvres of an aeroplane and an airship. Tethered balloons had been used for observing

9 TNA WO 279/39, Report on Army Manoeuvres 1910, p.63.

The airship BETA. (Alamy)

troop movements and artillery fire as early as 1879, and at the 1904 manoeuvres the defenders' balloon section had revealed the sites of the enemy's coastal landing, but this was the first use of an airship. Britain was well behind France and Germany in terms of using airships for military purposes. It was not until 1902 that Colonel Templer of the Balloon Section was given the go-ahead to carry out experiments with airships and not for another eight years would one appear at manoeuvres. BETA, completed in May 1910, carried a crew of three, could travel at 35 miles per hour and carried enough fuel for an eight hour journey. Her envelope had a volume of 35,000 cubic feet and the two propellers, wooden and two-bladed, were driven by a 35 horse-power Green engine.[10]

BETA had been used for aerial reconnaissance by the directing staff at the 1910 Aldershot Command inter-divisional manoeuvres but for the army manoeuvres it was placed under the command of the Red cavalry commander, Garrett, for six days and performed well, locating one of the two Blue infantry divisions on the first day, although she was grounded by strong winds on one day, and on another the dew from her envelope made her too heavy to fly until the morning sun evaporated it. In all, BETA flew over 700 miles and was operational on four of the five days. In order to pass aerial reconnaissance rapidly on to the Red cavalry, messages were dropped at marked spots near the front line. The first occasion on which troops had seen an airship on manoeuvres was also the occasion for the first use of anti-aircraft fire. Indeed, the instructions to umpires had anticipated this development, issuing guidelines for assessing its effectiveness.

10 G. Whale, *British Airships: Past, Present and Future*, pp.34-35.

Of less value in 1910, but greater significance in the long term, an aeroplane was employed by the Red army, though as the Air Battalion did not yet have sufficient flying experience to use its own aeroplanes, two Royal Artillery officers, Captain Bertram Dickson and Lieutenant Lancelot Gibbs, were given permission to use a privately owned Bristol Boxkite biplane, in a private capacity. The aeroplane was at first ignored, but on the second day it was placed under the command of the Red Force's cavalry. Dickson made a successful early morning reconnaissance on 21 September and successfully reported the location of the enemy's scouts.[11] When he landed at Codford St Mary to telephone his sighting, Dickson's aeroplane was surrounded by Blue troops who captured him before the umpires let him fly back to Red GHQ.[12]

The 1910 manoeuvres clearly showed the potential of aircraft. Garrett concluded that 'both airships and aeroplanes will have a very great value in the immediate future in war for reconnoitring purpose', though he did recommend that trained observers were needed as passengers and that the craft should be operated by soldiers rather than civilians.[13] It has suited some historians to portray British generals as technophobes who failed to recognise the way that aircraft would transform warfare, but Andrew Whitmarsh has conclusively demonstrated that, if we judge them by their actions rather than second-hand versions of what they supposedly said, they were well aware of the potential of aircraft and, having been comparatively slow to embrace the use of aircraft, the British Army was keen to make up for lost time. The Chief of the Imperial General Staff, General Sir William Nicholson, noted in February 1911 that 'In view of the fact that aircraft will undoubtedly be used in the next war… we cannot afford to delay in the matter.'[14]

Following the experience of the 1910 manoeuvres, the Conference of General Staff Officers at the Staff College in January 1911 devoted a great deal of time to a discussion of the new dimension aircraft would add to warfare. There was agreement that surprise might be hard to achieve in future, and that troops needed to be trained to march and operate after dark in order to avoid detection. *The Memorandum on Army Training* (1910) also demonstrated the army's increasing acceptance of the potential of aviation: 'It is probable that considerable developments will take place in the immediate future in the direction of reconnaissance by air craft in close association with cavalry… It is important for cavalry officers as well as commanders and staff officers to keep abreast of this movement.'[15] The July 1912 edition of the *Field Service Regulations* made reference to anti-aircraft fire, the need to conceal ground troops and the use of aircraft in tactical and strategic reconnaissance, especially in conjunction with cavalry.

11 A. Whitmarsh, 'British Army Manoeuvres and the Development of Military Aviation, 1910-1913', *War in History*, p.330.
12 www.bertram-dickson.com/aviators (accessed 30/8/2015).
13 TNA WO 279/39, Report on Army Manoeuvres 1910, p.182.
14 A. Whitmarsh, op. cit., p.326.
15 Ibid., p.332.

The cancellation of the 1911 army manoeuvres, in which it had been intended to employ four to six aeroplanes and two airships, denied the high command the opportunity to build on the success of aircraft in 1910; nevertheless, important work was done by the Air Battalion and then the Royal Flying Corps in 1911 and 1912 in terms of co-operation between aircraft and ground troops. Number 3 Squadron carried out tests and training, including air-to-ground and ground-to-air signalling, aerial photography and even night time reconnaissance. Trials were also carried out in observing artillery fire from aircraft on Salisbury Plain in co-operation with 3rd Division, whose commander, Rawlinson, took 'an intense interest' in the Royal Engineers' attempts to camouflage entrenchments from aerial observation.[16] By the time of the 1912 manoeuvres, the British were making up lost ground on the Germans and the acknowledged leaders in the field of aviation, the French. This time 15 aeroplanes would be involved as well as an airship on each side. For the first time, the aerial dimension was specifically mentioned in the appreciations prepared for each side, and the performance of the Royal Flying Corps would be subjected to unprecedented scrutiny.

By 1912, Britain's international position had been transformed from a decade earlier, and army manoeuvres reflected this change. James Grierson was unusual in stating as early as 1897 that 'We must go for the Germans and that right soon or they will go for us later.'[17] As far as most British officers were concerned in 1902, the enemy Britain would be facing in the event of war could equally have been Russia or Germany, whereas by 1912 it was quite clear that the purpose of their manoeuvres was to prepare the British army for a continental war in alliance with France against Germany. The signing of the Entente Cordiale with France (April 1904), along with the later Anglo-Russian Entente (August 1907) reflected increasing British concern at the growing threat from Wilhelmine Germany, which had embarked on an ambitious programme of naval expansion that appeared to challenge British imperial hegemony, as well as a shift away from an Asian policy dominated by fear of Russian designs on India.

The starting point for the so-called military conversations between France and Britain is often taken to be the first Morocco Crisis (1905-6), in which British diplomatic support was instrumental in enabling France to face down German threats and which was followed by the two countries moving closer together, a process in which generals rather than politicians set the pace. However, the British were already thinking in terms of a military connection to France even before the arrival of Kaiser Wilhelm in Tangier on 31 March 1905. The Director of Military Operations at the War Office, Grierson, had visited Paris earlier that month and assured French officials that Britain would fight on France's side in a future Franco-German war. Even before his visit to Paris, Grierson was conducting a war game which postulated a German invasion of France through Belgium, encountering joint Anglo-French resistance.

16 Ibid., p.334.
17 D. Macdiarmid, *The Life of Lieut. General Sir James Moncrieff Grierson*, p.133.

This war game merits detailed consideration, both as an example of the new seriousness of purpose regarding exercises and war planning exhibited by the British high command from 1902 onwards and because of its close relationship to the way events turned out in 1914.[18] It is unsurprising that Grierson should have been the moving force behind the war game, given his growing conviction that Britain needed to prepare for a continental war to resist German ambitions and his extensive experience of the German Army, which had long made more extensive use of war games than any other. The Prussian General Staff's use of *Kriegspiel* was seen as a major contributing factor in the country's victories in 1866 and 1870-71, to an extent that was probably exaggerated, but there is no doubt that war games, whether played on a board with model soldiers, conducted via tactical problems involving maps and written orders and responses or done in the form of staff rides, which took place outdoors, played a central part in German war planning.[19] This was especially true after 1876, when General Julius von Verdy du Vernois improved the original *Kriegspiel* by reducing the number of rules and making the game more realistic by giving experienced officers, acting as umpires, more power to adjudicate on combat results rather than relying on dice and combat tables.[20]

The British may not have embraced the use of war games in the systematic way the Germans did, but even before 1905 there had been tentative attempts to popularise war gaming in Britain such as Lt. Henry Chamberlain's *New Game of Invasion* (1888) and Spencer Wilkinson's *War Office Rules for the Conduct of the War-Game on a Map* (1896). Wilkinson, an enthusiastic but influential amateur who became Oxford University's first Chichele Professor of Military History in 1909, wrote of war games that 'Probably no form of military study is more useful if properly conducted, as certainly none is so liable to be misused,' and argued that they were preferable to manoeuvres, which 'like war itself, is too costly to be attainable except on rare conditions'. He observed that 'the only difference from actual war is the absence of danger, of fatigue, of responsibility, and of the friction involved in maintaining discipline.'

Grierson's War Game took place between January and May 1905, and was conducted with a modified strategic version of Wilkinson's 1896 War Game rules. It took place in real time and Grierson acted as the Chief Umpire, assisted by members of his staff. Three commanders were involved: Colonel William Robertson, Assistant Director of Military Operations, commanding the Germans, Colonel Charles Callwell, another ADMO, commanding the British and Major Arthur Lynden-Bell the Belgians. All three of these officers would have important roles to play in 1914, Robertson as QMG of the BEF and Lynden-Bell as his Assistant, while Callwell served in the War Office

18 TNA WO 33/364, Record of a Strategic War Game, 1905.
19 Bucholz, op. cit., pp 85-92.
20 For the 1905 War Game, and British wargaming before 1914 in general, see Christopher Yi-Han Choy, British War Gaming, 1870-1914 (King's College London MA in Conflict Studies, 2013).

as DMO. They were assisted in the War Game by junior staff officers from Military Operations and Foreign Intelligence, many of whom would find themselves involved in the fighting on Belgian soil nine years later. To take one example, Major D.J.M. Fasson of the Royal Artillery, who acted as the German chief of staff for the purposes of the War Game, found himself confronted with a very real German army when he commanded XXII Brigade as part of 7th Division, which landed at Zeebrugge in October 1914. The division was originally intended to assist in the defence of Antwerp, but after its fall they marched south to Ypres.

The decision was taken not to represent the French, as all the action would take place on Belgian soil. The game depended for its efficacy on the excellence of German cycling maps, which included vital details such as the lesser known by-roads in the south of the country. The 'General Idea' assumed that war had broken out between France and Germany on 1 January 1905, and that after two months of failing to break through the French defences between Sedan and Belfort, the Germans had launched an outflanking movement through Belgium. This infringement of Belgian neutrality had brought Britain into the war. Grierson and his planners were highly sceptical of the Belgian army's ability to garrison their fortresses while also maintaining a presence in the field, and they had identified serious flaws in the Belgian fortress defences. The 'General Idea' proposed that the Germans would seek to destroy the Belgian field army before dealing with the fortresses.

A force of six Army Corps, three cavalry divisions and two Reserve Army Corps was employed in the German invasion of Belgium. The bulk of this force advanced south of the Meuse while one infantry and one cavalry division launched a feint north of Liège. The Belgians fell back in the face of this offensive, redeploying to garrison Antwerp and secure the landing ports for the British. Robertson, commanding the Germans, advanced across the Ourthe towards Dinant, confident in the belief that the British were still disembarking and that the Belgians would do nothing until this process was complete, but an Anglo-Belgian counter-attack on the 30th day of mobilisation brought the game to an end. Grierson ruled in favour of a German victory, as he believed that even if the counter-attack had been successful, their advance 'could not have been materially interfered with, until the arrival of the greater portion of British troops.'

The war game featured a significant disagreement between the putative allies, with 'the British Commander taking the purely military, and the Belgian the political-military point of view.'[21] Callwell argued that the Belgians should have deployed in the north from the very start, basing their defence on Antwerp and the landing ports, whereas Lynden-Bell insisted that the Belgians could not be seen to be surrendering their southern provinces so easily, and Grierson sided with this decision, both for reasons of domestic political considerations and because of the effect on Belgium's allies if she was perceived not to be defending her neutrality with sufficient vigour.

21 TNA WO 33/364, Record of a Strategic War Game, 1905, p.153.

The proposed British force – three army corps and three cavalry brigades – was intended to take seven days to ship across to Belgium, based on forty-two transport vessels, but it was established during the course of the game that with only twenty-two vessels available ten days after mobilisation, and the rest only available after seventeen days, it would take 34 days for the entire force to be transported.

Three important lessons were learnt –or, at least, confirmed – by Grierson and other war planners as a result of the 1905 War Game: first, that France was likely to lose to Germany on her own, second, that the current mobilisation plans were unlikely to enable the BEF to reach France quickly enough to make a difference, and third, that the British army was too small to be able to intervene to save France and Belgium. It was clear that there would be little chance of stopping the German turning movement unless the British forces arrived on the scene quickly and in considerable strength.[22] These conclusions formed the basis of the General Staff's responses to the three questions posed by Prime Minister Arthur Balfour later in 1905:

1. What might Germany – or France – gain by a violation of Belgian neutrality?
2. Could Belgium be expected to offer meaningful resistance?
3. How long would it take for two British corps to arrive?[23]

The General Staff's responses to these questions – Germany was likely to seek a strategic advantage by an invasion of Belgium, which could not offer effective opposition by herself, and a British deployment would take at least twenty-three days even if all went well – would be highly influential in British planning for the next nine years. It is tempting to describe the War Game as providing an 'uncanny' prediction of later events, but in reality Grierson and his colleagues came to an obvious set of conclusions based on the limited number of options for the British to choose from. Sir John French, who would ultimately command the expeditionary force which would be despatched to assist Belgium when war came, wrote to Grierson to express his appreciation: 'Thank you *very very* much for the account of your War Game. I've nearly finished it *once* and shall probably start at it a second time before I go to bed tonight! It is most interesting and instructive.'[24]

As an example of the evolving strategy which was evolving in the aftermath of the War Game, the War Office memorandum written in October 1905 by Callwell entitled 'British Military Action in Case of War with Germany' stated that 'the most useful purpose to which (the army) could be put would be to give support to the French armies in the field.'[25]

22 Ibid.
23 Jason Tomes, *Balfour and Foreign Policy: The International Thought of a Conservative Statesman,* p.135.
24 D. Macdiarmid, op. cit., p.212.
25 J. McDermott, 'The Revolution in British Military Thinking from the Boer War to the Moroccan Crisis' in *The War Plans of the Great Powers, 1880-1914* ed. P. Kennedy, p.110.

Grierson spoke to the French Military Attaché, Colonel Victor Huguet, in London in December 1905, at the peak of the Morocco Crisis, when Huguet expressed his fears of an imminent German attack and asked if the British had ever given any thought to operations in Belgium. Two meetings of the CID were crucial to the process; on 6 January 1906 the Army strategy (prepared by Grierson but presented by Sir John French) was accepted, and the idea of joint planning with France approved, and on 19 January, with no Admiralty representative present, it was left to Grierson and French, Lord Esher and Sir George Clarke, Secretary of the CID, to consider a plan requiring an expeditionary force to entrain at French channel ports on the twelfth day of mobilisation.

With Foreign Office approval Grierson then opened a series of conversations with the French that would continue up to the outbreak of war. Grierson and other staff officers embraced with enthusiasm the idea of war planning with Germany as the enemy. The opportunity to have a clearly defined role as the French left wing in a continental war would enhance the Army's status and establish its independence both from India and from the Admiralty.

The key figures in this process were Grierson and Henry Wilson, who became DMO in August 1910. Grierson's visit to the French manoeuvres of 1906, accompanying Sir John French, and the friendly reception they received there, was a crucial development in the growing understanding between the two countries, though this relationship was largely dependent on a few Francophile individuals at the top of the British Army and took place without the active participation of ministers. Most of the Cabinet were unaware of the conversations until 1911, and those who were in on the secret, above all Haldane and the Foreign Secretary, Sir Edward Grey, were able to disavow knowledge of them when necessary. When the German Ambassador complained of Huguet's regular visits to the War Office, Haldane replied 'This is the first I've heard of it, but after all what harm is there in the two Staffs getting together? We are in the process of reorganising the Army and if the French choose to help us, why shouldn't we take advantage of it?' He did, though, ask Huguet to use the back entrance in future.[26]

Two points were stressed repeatedly, that the conversations in no way represented a binding commitment on Britain's part, and that British involvement in a continental war was contingent on a German violation of Belgian neutrality, which would justify a British intervention under the 1839 Treaty of London. For this reason, as well as talking to the French, Grierson was keen to find out Belgian intentions, instructing the Military Attaché in Brussels, Colonel Nathaniel Barnardiston, to open a dialogue with the Chief of the Belgian General Staff, Major-General G.E.V. Ducarne.[27]

26 General V. Huguet, *Britain and the War: A French Indictment,* p.16.
27 There is much debate over the scope and significance of these discussions. For a highly partial view from 1915, see Alexander Fuehr, *The Neutrality of Belgium.*

Grierson and French at the 1906 French manoeuvres. (Author's collection)

Once again, it was stressed that these talks, which commenced in January 1906, were provisional and non-committal on Britain's part. Grierson instructed Barnardiston: 'You tell the Chief of Staff what we are prepared to put in the field in this case, 4 Cavalry Brigades, 2 Army Corps, and a division of mounted infantry… The total numbers will be about 105,000.'[28] Barnardiston told Ducarne that in the event of Belgium being attacked, these troops would be landed at the French ports of Calais, Boulogne, Dieppe and Le Havre, and then transported to Belgium by rail. Antwerp was considered and rejected as an alternative landing site, probably as a result of lessons learnt during the War Game, though Grierson stipulated that the base of operations could change to Antwerp later when command of the sea was assured. In return, Ducarne informed Barnardiston of the defensive positions which the Belgian Army of 100,000 men would take up in the event of a German invasion. In later conversations, topics included the unlikelihood of Dutch support, the strength of the fortresses of Liège and Namur and the importance of the Belgian railway system in speeding up the British deployment.

As a proudly neutral country, Belgium needed careful handling. Barnardiston's successor, Colonel Tom Bridges, carried out similarly confidential conversations with Ducarne's successor, General Harry Jungbluth in April 1912, assuring him that Britain could commit 150,000 men 'at the decisive point and time', but this only

28 M. Connelly, 'Lieutenant-General Sir James Grierson' in S. Jones (ed.), *Stemming the Tide*, p.138.

succeeded in alarming the Belgian War Minister, General Augustin Michel, who had no desire to see his country dragged into a war and was suspicious of British interest in Antwerp and the Scheldt.[29]

The Belgian question continued to preoccupy British military planners, prominent among them Wilson, by now Commandant of the Staff College. In November 1908 Wilson set his senior division an exercise, the 'Belgian Scheme', which entailed a British continental deployment to assist Belgium.[30] It started with the warning that there were many sets of circumstances which could make such a deployment necessary, and for the purposes of the exercise certain actions by the British government were assumed, 'but it is not to be assumed that these policies in any way represent present day conditions'. The exercise assumed a deterioration in Franco-German relations. Germany was attacking France in order to break up the entente with Britain, and would 'violate Belgian neutrality if such a step will assist her ultimate designs against England'.

The students were divided into seven syndicates, each one consisting of five or six officers, and they were instructed to provide for the Cabinet a memorandum containing the General Staff's views 'as to the most effective means of employing the British Expeditionary Force, when its mobilisation is completed'. The exercise was set on Monday 23 November and the students had until the following Saturday to complete it. Such was the sensitivity of the subject matter that it led to questions being asked in the House – 'whether we were to be permitted to hatch malicious plots against the harmless, peace-loving Germans', according to one member of Wilson's staff.[31] When the exercise was repeated in November 1909 there was no reference in the instructions to a German violation of Belgian neutrality. This time it was made clear that the exercise 'and all work connected therewith, must be regarded as SECRET'.[32]

Wilson's 'Belgian Scheme' shared with Grierson's War Game a new determination to anticipate the circumstances which might see Britain engaged in a continental war. In the same way, Grierson's visit to the Belgian manoeuvres in August 1906 must be seen in the context of a desire to gauge the capabilities of the Belgian army in the event of war, and in this respect what he saw was not encouraging: 'I do not think that the Belgian army is in an effective state of preparation for war.'[33] This helped to persuade Grierson that in the event of a continental war any British force should cross to French rather than Belgian ports and deploy on the Belgian right.[34] He was also able to meet General Ducarne at Compiègne during the French manoeuvres.

29 S. Williamson, 'Joffre Reshapes French Strategy 1911-1913' in *The War Plans of the Great Powers, 1880-1914* ed. P. Kennedy, pp.141-142.
30 K. Jeffrey, *Field Marshal Sir Henry Wilson: A Political Soldier*, p.73.
31 General Sir G. de S. Barrow, *The Fire of Life*, p.115.
32 K. Jeffrey, op. cit., p.73.
33 D. Macdiarmid, op. cit., p.219.
34 M. Connelly, op. cit., pp.138-9.

Army manoeuvres played an important role in bringing the two sides together and forming relationships between key individuals on each side. The most important connection was between Ferdinand Foch and Henry Wilson, who was a regular attender of French staff tours and exercises, and even attended the wedding of Foch's daughter Marie in October 1910. He was also invited to, but did not attend, the wedding of the younger daughter Anne in October 1912. In all, Wilson paid 11 visits to the French between December 1909 and May 1914.[35] Sir John French attended French manoeuvres in 1906, 1908 and 1911, Grierson in 1906 and 1909. The positive general impression Grierson and French took away with them in 1906 did much to strengthen the British commitment to the Entente. As war approached, links between the two high commands grew ever stronger, especially after the Second Morocco Crisis (1911) reiterated the German threat. Foch attended the 1912 British manoeuvres and General de Castelnau those of 1913, while Joffre, Chief of the French General Staff from July 1911, was due to do so the following year. General Jungbluth was also invited to attend the 1912 British manoeuvres but declined, probably because of fears that his presence might be interpreted in Berlin as a sign of a growing understanding between the general staffs of Belgium and Britain.[36]

The French had concerns over the British alliance, worrying about the lack of a concrete commitment from the British and the pacifist tendencies of the left of the Liberal Party. The Liberal government's difficulties in Ulster, typified by the Curragh Incident early in 1914, also troubled the French. After the war, Joffre admitted that 'Personally, I was convinced that they would come, but in the end there was no formal commitment on their part. There were only studies on embarking and debarking and on the positions that would be reserved for their troops.'[37]

Nevertheless, even if the British avoided any binding commitments to France, important work was done in preparing for the moment when war came. The original proposal of British assistance made in 1906 – an expeditionary force of six infantry divisions and a cavalry brigade to be despatched to one or other of the Channel ports and thence to Belgium – remained in place, despite changes of personnel on both sides, for the next eight years. An important breakthrough was achieved in 1908 when the transport timetables for the British Army were fixed on the same plan as for the French, and the schemes agreed in 1911 would be implemented essentially intact three years later, ensuring the smooth process by which the BEF arrived at Maubeuge in August 1914.

Another significant step in this process was the meeting in August 1911 of a sub-committee of the CID at which the Army and Navy presented their differing strategies for a war between France and Germany. The Admiralty's case for a close blockade of Germany and amphibious landings on the Baltic coast was presented poorly by

35 E. Greenhalgh, *Foch in Command: The Forging of a First World War General*, p.10.
36 Baron Beyens, trans. P.V. Cohn, *Germany before the War*, p.319.
37 R. Doughty, 'French Strategy in 1914: Joffre's Own', *The Journal of Military History*, p.436.

the First Sea Lord, Admiral Sir Arthur Wilson, and met with disdain from the politicians present, whereas Henry Wilson for the Army made a convincing case for the commitment of an expeditionary force to the continent. The significance of one meeting should not be overstated, and it has been argued persuasively that the Navy's 'defeat' was not as clear cut as has usually been presented, but even if Sir Arthur Wilson's strategy was actually more coherent and far-sighted than most historians have been prepared to admit, the perception that the Army had eclipsed the Navy played an important part in the Liberal government's acceptance of the principle of a 'continental commitment'.[38]

This shift to a continental strategy roughly coincided with important reforms which enhanced Britain's military capability. Examining the manoeuvres at the end of the decade, one gets a clear sense of progress from the fumbling efforts in the immediate aftermath of the Boer War, and this reflected a growing professionalism which was transforming the British Army; in his study of the role of military intelligence in British rule over India, James Hevia writes of 'a sea change in the organization and planning of warfare in Great Britain' from around 1903. A serious-minded and dedicated group of officers, many of whom had made their name in Sir Henry Brackenbury's Intelligence Department, were keen to employ all means at their disposal, whether war games, staff rides, manoeuvres or intelligence gathering, as well as new and improved staff manuals, field service manuals and army handbooks, to prepare for war. Hevia argues that the period witnessed 'a wholly new centralized structure of military oversight in Britain, one fully in the hands of a professional elite'.[39] The creation of a General Staff in February 1904 (succeeded by the Imperial General Staff in 1909) was one of the many developments which enabled this transformation, even if this was not yet a General Staff with the prestige and ethos of the German model. The Staff College at Camberley, under a succession of energetic Commandants, Henry Rawlinson, Henry Wilson and William Robertson, increasingly attracted ambitious officers but when war came there were still only 447 officers with the magical initials *psc* (passed Staff College) in the entire British Army.

Along with the creation of a General Staff, the other crucial development was the formation of a new Liberal government in December 1905, or more specifically the arrival at the War Office of a new Secretary of State for War, Richard Haldane, whose far-reaching reforms essentially created the army which would go to war in August 1914. The home army was reorganised in January 1907 into six infantry divisions and a large cavalry division, which would together provide a 'striking force' for the continental commitment. The 1907 Territorial and Reserve Act, which provided for 14 territorial infantry divisions and 14 Territorial cavalry brigades, was designed

38 See the important re-evaluation in D. Morgan-Owen, 'Cooked up in the Dinner Hour? Sir Arthur Wilson's War Plan, Reconsidered', *The English Historical Review*.

39 J. Hevia, *The Imperial Security State: British Colonial Knowledge and Empire-Building in Asia*, p.167.

to produce a force which would relieve the Regular Army of home defence duties, allowing it to be despatched to the continent. In all of these reforms Haldane was assisted by a cadre of able and ambitious generals, notably Sir Douglas Haig (Director of Military Training 1906-7, then Director of Staff Duties 1907-9), Sir James Grierson (Director of Military Operations 1904-6) and Sir Henry Wilson (Director of Military Operations 1910-14). Grierson and Haig would play central roles in the last two sets of manoeuvres before the outbreak of war.

With the high command now increasingly thinking in terms of a future war being fought in Belgium or northern France, questions were once more raised about the desirability of training on land that would provide different challenges from the open spaces of Salisbury Plain. Writing in the *United Service Magazine* for 1905, 'Foresight' questioned the very notion of a special manoeuvre area. 'Still we go with the same old training year after year, day after day… and which the taught soon know by heart, and of which the teacher knows every bush.'[40] The decision to stage the 1912 manoeuvres in East Anglia was therefore widely welcomed. As well as taking place in unfamiliar terrain, they would be on a different level to those of previous years because of the numbers involved, the choice of location and the importance given to administrative, supply and transport issues, as well as the first significant use of airships and aeroplanes. They would also pit the two rising stars of the British Army against each other, two men who could reasonably expect to have leading roles in the expeditionary force if and when war came, Haig and Grierson.

Lieutenant-General Sir Douglas Haig had made his name as French's chief staff officer in South Africa and was the army's leading expert on the cavalry, while Lieutenant-General Sir James Grierson was its acknowledged authority on manoeuvres – he was chosen to pen the entry on the subject in the 1911 Encyclopaedia Britannica and he had extensive foreign experience, having attended numerous German manoeuvres in his capacity as Military Attaché in Berlin between 1896 and 1900 and, more recently, those of France and Belgium. In addition, he was a gifted linguist, having been successful in examinations for the role of interpreter in French, Russian and German (he also spoke Italian and Hindi) and a prolific author on military matters, writing extensively on the armed forces of Russia, Turkey, Japan and Germany. Unusually in an army where cavalrymen were allegedly over-represented in the higher levels of command, Grierson was a Gunner.[41] After service in the Intelligence Department and in Berlin, he had seen action in South Africa as QMG under Lord Roberts, gaining invaluable staff experience in an army of 37,000 men.

40 T. Crawford, *Wiltshire and the Great War: Training the Empire's Soldiers*, p.17.

41 After Grierson the senior Gunner in the BEF in August 1914 was Brigadier General Sir Henry Horne. The myth of the domination of the BEF by cavalrymen was convincingly dealt with by J. Terraine, *The Smoke and the Fire, Myths and Anti-Myths of War 1861-1945*, pp.161-163.

Having played a vital part in the reforms associated with the Esher Report and with Lord Haldane, Grierson had commanded the 1st Division at Aldershot between 1906 and 1910 and had entertained hopes of being appointed to the Aldershot Command, the plum command in the peacetime army, in 1912. Instead, the appointment had gone to Haig, and Grierson had to content himself with Eastern Command, a far less challenging and prestigious post.[42]

Haig could not compete with Grierson in terms of intellect, but he was conscientious and serious-minded, an 'educated soldier' who had taken his studies at the Staff College seriously and who was far more receptive to new ideas and methods than purveyors of the standard image of him as the archetypal cavalry officer would have us believe.[43] He had been an energetic reformer while at the War Office and had contributed a number of important works to the ongoing debate over the role of cavalry, including *War and the Arme Blanche* (1910) and *German Influence on British Cavalry* (1911). Haig's most distinctive contribution to staff training came in the form of staff rides, which he made more use of than any of his contemporaries. These tactical exercises without troops, 'Tewts' as they would come to be known, were an important feature of his time as Chief of Staff in India – he carried out five staff rides in his three years there, each lasting a week.[44]

Given their differences in temperament and outlook, it is tempting to see Haig and Grierson as terrestrial equivalents of Jellicoe and Beatty – a simplistic dichotomy which has dominated too much thinking on the Royal Navy in the Great War – but they did have much in common, above all a deep-rooted seriousness of purpose when it came to their profession. Both men were lowland Scots, and shared a strong religious faith rooted in their Presbyterian upbringing, with Haig later describing Grierson in terms which could have been used for himself: 'A Scotsman to the finger-tips, he loved the simple religion of our Fathers.'[45]

The two men were friends as far as we can tell, the bachelor Grierson staying with the Haig family for Christmas 1912, three months after the manoeuvres. Haig's foreword to Macdiarmid's 1923 biography of Grierson certainly suggests warm feelings towards his old comrade. However, in most respects their personalities were very different. Grierson was a gregarious and popular individual much given to amateur dramatics and late nights in the Officers Mess, whereas Haig, if not entirely humourless, was a dour man who ate and drank in moderation and found small talk excruciating. It is not easy to imagine him cavorting around in a dress as Martha in a production of *Faust Fin de Siècle* as Grierson did at Aldershot in 1895. According to

42 For Grierson, see D. Macdiarmid, *The Life of Lieut. General Sir James Moncrieff Grierson.* Macdiarmid's biography was written with reference to Grierson's papers and diary, which were burnt by Grierson's family on Macdiarmid's death in 1954.

43 See G. Sheffield, *Douglas Haig: From the Somme to Victory* for the most balanced view of Haig available.

44 W. Reid, *Douglas Haig: Architect of Victory*, p.112.

45 S. Batten, 'The Rivals: Haig and Grierson', *The Douglas Haig Fellowship Records,* p.35.

Charteris, Haig 'viewed with some misgivings the growing tendency of the younger officers to seek their relaxation in dancing rather than more virile exercise.'[46]

However, whereas Haig was trim and fit and had a robust constitution despite childhood asthma, Grierson, always a bon viveur, was extremely corpulent; Edmonds, admittedly not the most reliable of sources, recalled that Grierson used to go red in the face when bending over, due to high blood pressure, and that his medical officer had issued Grierson's staff with penknives so that he could be bled when his blood pressure rose too high.[47] On his way to the Belgian and French Manoeuvres in September 1906, he had stayed at Marienbad, managing to lose 7¼ lbs during his visit. While there he dined with Haldane as well as the Prime Minister, Henry Campbell-Bannerman, and King Edward VII, who told him 'I know why you have come here. It is to get your figure down to Aldershot dimensions.'[48] Writing to Edmonds in August 1911, Haig expressed the opinion that 'we don't pay nearly enough attention to the physical fitness of the Staff Officers, especially at the War Office.'[49]

According to the panegyrical biography written by Duncan Macdiarmid in the early 1920s, Grierson in 1912 was 'a confident, happy, buoyant figure, cool and unperturbed, stationary at the centre of affairs, receiving reports, alert, with a comprehensive grip upon each situation as it developed, issuing the needful orders, the possessor, above all, of that subtle quality – personal magnetism.'[50] This last observation was not one that anyone was ever likely to make of Sir Douglas Haig, though he was certainly capable of commanding great loyalty from his subordinates. *The Manchester Guardian* fully expected the coming manoeuvres to show Grierson to be the better general – although Haig was praised for his appreciation of a situation, 'I find, however, that many knowing ones believe that the result of the manoeuvres will be to display General Grierson as a scientific officer of the very highest order, and that General Haig will suffer by the comparison.' The newspaper went on to make a perceptive observation on the two rivals' ability to work with their staff: 'General Haig is less tolerant of advice than his rival, and the value of the staff will therefore be more evident in the acts of General Grierson.'[51] The press would be present at the manoeuvres in greater numbers than ever before, with 39 journalists attending and meeting with Captain Stuart of the General Staff at 4:00 p.m. each day, when he would discuss that day's operations, distribute the previous day's official narrative and tip journalists off about the best place to see the action the next morning.

46 J. Charteris, *Field-Marshal Earl Haig*, p.66.
47 Liddell Hart conversation with Edmonds 5 Feb 1937, quoted in T. Travers, *The Killing Ground. The British Army, the Western Front and the Emergence of Modern Warfare 1900-1918*, p.14.
48 Macdiarmid, op. cit. p.218.
49 Douglas Haig (ed. Douglas Scott), *The Preparatory Prologue: Diaries and Letters 1861-1914,* p.61.
50 Macdiarmid, op. cit. pp.245-246.
51 *The Manchester Guardian*, 9 September 1912.

On paper the two sides were evenly-matched, each having two infantry divisions, a cavalry division and similar numbers of guns, machine guns and aircraft.[52] If anything, Grierson's Blue army had the edge, with an additional Territorial detachment of four battalions of the Liverpool Regiment giving Grierson a slight numerical advantage (23,564 to 21,392), although *The Times* warned that Territorials could hardly be expected to hold their own against trained Regulars.[53] Furthermore, the Territorial Detachment was a very makeshift unit – while all the infantry came from the same regiment, their support units were cobbled together from a variety of sources: the 2nd Northumbrian Brigade RFA, engineers from the East Anglian Division, the supply train from the Liverpool and Northumbrian Brigades and the medical units from the 2nd W. Lancs Field Ambulance.

In all respects other than numbers, Haig's Red force was clearly the stronger. He could call upon the two infantry divisions of his Aldershot Command and enjoyed the luxury of a settled staff used to working with each other, whereas Grierson's force was improvised from various Commands. The situation reminded Grierson of his observations on the 1896 German Manoeuvres when his detailed report for the Intelligence Division had contrasted von Waldersee's staff ('an intimate and happy family') with Prince Georg of Saxony's 'which was got together from all quarters and worked with endless friction'.[54]

The disparity was most obvious in the mounted arm; the Red division consisted of three brigades of regular cavalry, commanded by Major-General Edmund Allenby, the Inspector-General of Cavalry, whereas the two Blue brigades comprised a miscellaneous band of regular cavalry units, Mounted Infantry, cyclists and Yeomanry, under the command of a Colonel, Charles Briggs. *The Times* observed that the Blue cavalry were at a significant disadvantage, with the Yeomanry 'naturally being comparatively deficient in training and entirely incapable of offensive shock action.' After witnessing the effectiveness of aircraft in a reconnaissance role during divisional training, *The Times* was unwilling to predict which side would prevail, so evenly matched were they: 'Either of the rival commanders will have to make a great call on the endurance of his men if he hopes to achieve any sudden and unexpected result.'[55]

The manoeuvres would be overseen by a Director, none other than the Chief of the Imperial General Staff, Sir John French, while Bruce Hamilton (Red) and Horace Smith-Dorrien (Blue) were to be the Chief Umpires. The King would arrive from Balmoral on the second day, accompanied by Major-General William Robertson and the Secretary of State for War, Jack Seely, who had succeeded Haldane in 1913 when he became Lord Chancellor. Before Seely's appointment, Henry Wilson had

52 Red: 96 guns, 64 machine guns, 6 aeroplanes, 1 airship. Blue: 99 guns, 67 machine guns, 9 aeroplanes, 2 airships. TNA WO279/47, Report on Army Manoeuvres 1912.
53 *The Times*, 17 September 1912.
54 Macdiarmid, op. cit., p.125.
55 *The Times*, 17 September 1912.

written in his diary, 'I suppose we shall get Jack Seely. Ye gods!' Seely was widely seen as a political lightweight, whose distinguished service as a Captain in the Imperial Yeomanry in South Africa was not the recommendation for the job it might appear. As his grandson, Brough Scott, has noted, 'A madly enthusiastic yeomanry colonel was exactly what Wilson and the other generals did not want.'[56]

The manoeuvres would be centred on Cambridge, with the decisive action taking place in southern Cambridgeshire, where the woods and gently rolling hills presented a very different challenge to the more familiar training ground of Salisbury Plain. This was not the flat, featureless terrain generally associated with East Anglia. A guide-book of the time described the area as 'good, agricultural countryside, slightly undulating, sufficiently wooded, sprinkled with sleepy little market towns, and abounding in villages of the old-world type'.[57] A sizeable zone from Huntingdon to the Wash was marked 'impassable for troops' on manoeuvre maps, and as the press commented after weeks of incessant rain, 'The Fens were thus marked long before the wet weather and the floods, but now this description is singularly appropriate.'[58]

The 'General Idea' issued to participants imagined that Red forces under Haig had crossed the frontier (the coastline of the Eastern counties) into Blueland from northern Norfolk and were moving rapidly southward. It was the task of the Blue commander, Grierson, to stop Red from taking the capital, London. Two Blue infantry divisions were en route by railway from Aldershot and Salisbury and Blue mounted troops were located west of Saffron Walden, while the Territorials, about a third of a division strong, were at Cambridge. One Territorial, a member of the 6th Rifle Battalion, Liverpool Regiment, told a newspaper reporter that:

> It is a great honour for us, and, of course, we are all eager to show our mettle, and be worthy of what is expected of us… … we are quite excited about it. It will be so novel for many of us though some at present in the battalion served in the Boer War.[59]

An invasion of East Anglia was a plausible scenario in the event of war with Germany: numerous contemporary examples of invasion literature entailed German invasions of the east coast of England; John Tregellis' *Britain Invaded* (1906) starts with the Germans landing five army corps at Hull, Boston, Cromer, Lowestoft and Frinton, while Erskine Childers' *The Riddle of the Sands* (1903) features plans for a German invasion of the east coast by a secret fleet of barges assembling off the Frisian Islands. East Anglia was also the setting for the early chapters of William Le Queux's

56 B. Scott, *Galloper Jack*, p.134.
57 R. Wellbye, *Road Touring in Eastern England*, p.33.
58 *The Cambridge Daily News*, 7 September 1912.
59 *The Cambridge Daily News*, 5 September 1912. This is a useful reminder that the Territorials included some South Africa veterans in their ranks.

sensationalist novel *The Invasion of 1910*. This had originally been commissioned by Alfred Harmsworth as a serial in *The Daily Mail* from 19 March, 1906. The shrewd decision to rewrite the story to feature towns and villages with high *Daily Mail* readership boosted sales and made Le Queux a great deal of money, as well as focusing attention on the issue of Britain's preparedness in the face of the perceived threat from Germany – the author's avowed aim in writing the book was to 'bring home to the British public vividly and forcibly what really would occur were an enemy suddenly to appear in our midst.' Over one million copies of the book version were sold and there were translations into twenty-seven languages. *The Invasion of 1910* involves the by now familiar landing on the coast of East Anglia, followed by a German advance on London and a battle at Royston on the border between Hertfordshire and Cambridgeshire, close to where the climactic 'battle' of the manoeuvres would take place in 1912.[60]

60 I. Clarke, *Voices Prophesying War: Future Wars 1763-3749*, p.122-123. For a valuable collection of resources on invasion literature see www.theRiddleoftheSands.com. (accessed 30/8/2015).

# 4

# Haig v Grierson: The 1912 Manoeuvres

By Sunday 15 September 50,000 troops were in place ready for the manoeuvres to commence. Many had already been in the area for some time, taking part in large-scale cavalry manoeuvres during the first week of September, followed by the inter-divisional manoeuvres in the second week. The weather in East Anglia had been wet for several weeks, with local newspapers reporting 'A Record August. Only Five Rainless Days.'[1] It was the wettest and coldest August on record, with widespread flooding in Suffolk and Norfolk. At one point the weather threatened to wreck the Army Manoeuvres for the second year in succession; George V's Private Secretary, Lord Stamfordham, sent a telegram from Balmoral on 27 August to the Secretary of State for War, Jack Seely, to convey the King's concern that the manoeuvres might have to be cancelled – 'If the weather does not improve will not living under canvas be detrimental to the troops? Are the newspaper accounts of the conditions of the troops living in camp and the Aldershot district correct and is there any idea of abandoning the manoeuvres?'[2] Seely's response was that the conditions were indeed bad but that it would not take a great deal of fine weather for the ground to dry out, and so it proved. The most significant set of manoeuvres yet attempted by the British Army would go ahead as planned.

Nevertheless, the heavy rain persisted. One Coldstream Guards officer noted 'incessant rain all night and day' in his diary for 11 September,[3] and early morning mists were to play a significant role, as they would in Flanders in August 1914. With the Blue force concentrating somewhere on his right (northern) flank, Haig was reluctant to advance too far towards London, so in the absence of further information he decided on an immediate but cautious advance. If Grierson concentrated his forces north of London Haig would advance to force a battle, and if he pushed forward from Cambridge, Haig would attack Blue's right (southern) flank to drive him back

1 *The Cambridge Daily News*, 3 September 1912.
2 The papers of John Edward Bernard Seely, Lord Mottistone, 17 58-61.
3 Diary of Capt. J.H. Brocklehurst, Coldstream Guards Archive, Wellington Barracks.

Major-General Lawson and the 2nd Division staff at Eriswell, September 1912. (Saanich Archives)

towards the Fens. If, however, Blue advanced a long way from Cambridge, Haig would concentrate his whole force against Blue's left flank in order to cut his line of communications to the west.

Grierson thought he knew which way Haig would come in order to advance on London, thanks to the fen country along the western boundary of the manoeuvre area and the difficult terrain between Bury St Edmunds and Newmarket. He considered moving south in order to cover London and take advantage of the favourable ground, but in the end he decided that this option would surrender the fortified pivot of manoeuvre he had in Cambridge and might be difficult to carry off when fighting was imminent. Instead, he ordered the Territorials to dig in around Cambridge while he sought to ascertain Red's direction of march by dispatching the cavalry and aircraft to reconnoitre around Newmarket and Mildenhall. His two infantry divisions, the 3rd under Major-General Henry Rawlinson and the 4th under Major-General Thomas Snow, had only just detrained and would not yet be able to move forward. According to the report compiled for the Canadian Ministry of Militia and Defence, the concentration by rail of Grierson's force was the outstanding feature of the opening phase of operations: 'Without any appreciable delays a succession of troop trains poured into the railways stations at Bedford, Gamlingay, Hitchin and Milbrook, discharging

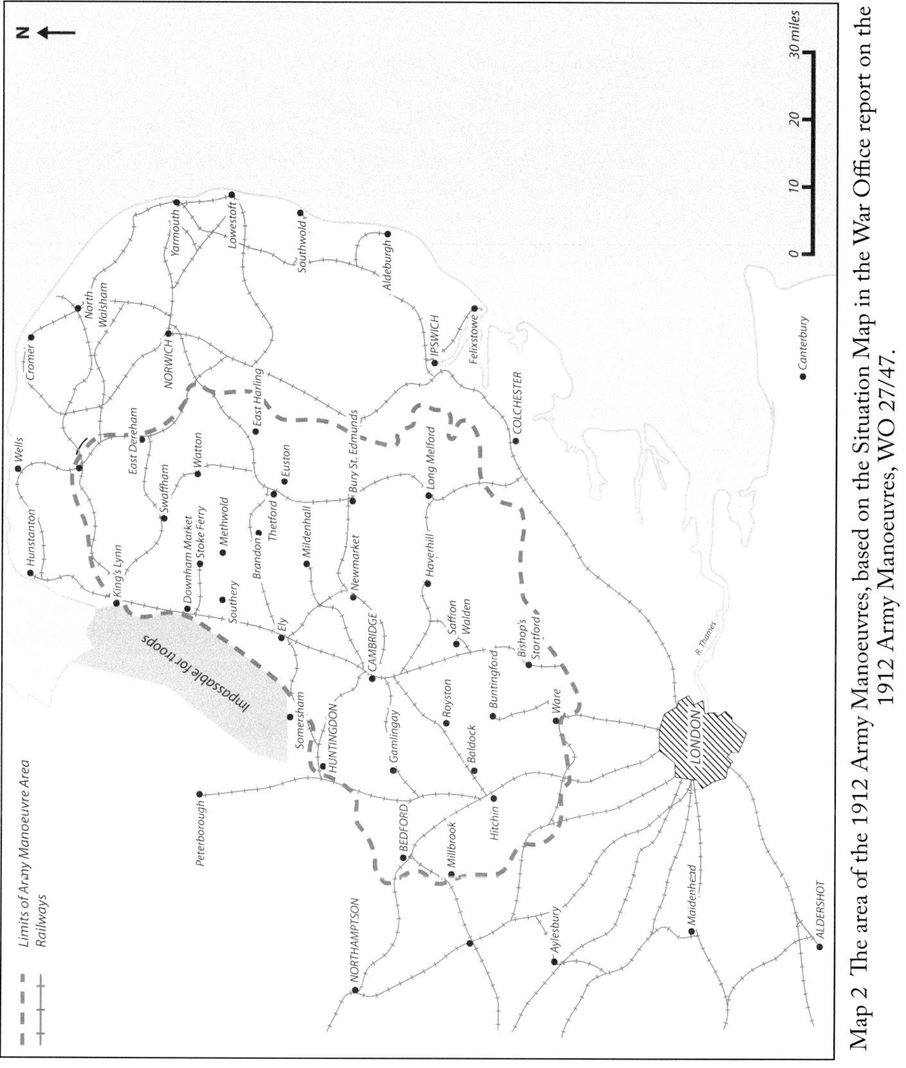


Map 2 The area of the 1912 Army Manoeuvres, based on the Situation Map in the War Office report on the 1912 Army Manoeuvres, WO 27/47.

troops of all arms and administrative units, who at once set themselves in motion towards the concentration bivouacs allotted to them.'[4] Once the direction of Red's march could be ascertained, then the whole force would advance to battle. Operations would commence at 6.00: a.m. the next day.[5]

## Day 1: Tuesday 16 September

From the outset both sides made extensive use of aerial reconnaissance. Each had one airship. Haig was assigned DELTA, built earlier in the year and capable of 44 miles an hour, but it broke down en route from Farnborough over North London, and so was replaced by BETA, now re-equipped with a 45 horse-power Clerget engine. Grierson could call upon GAMMA, completed in 1910, a much bigger ship with an envelope of 101,000 cubic feet capacity to BETA's 32,000 cubic feet. Both GAMMA and DELTA had been equipped with wireless apparatus, but BETA was too small to carry the installation, with the result that only the Blue army could call upon wireless communication, though this was still a one-way process as airships could not yet receive wireless messages, only transmit them. In addition to the airships, Blue could call upon nine aeroplanes of various types, including BE (Blériot Experimental), Breguet and Maurice Farman aircraft, while the Red squadron had six.

From the start, it was clear that aircraft would play a far bigger role than in the 1910 manoeuvres. Having witnessed the use of aeroplanes at Indian Army manoeuvres in 1911, Haig was keen to employ this new means of reconnaissance to the full.[6] By noon on the first day, he was convinced by aerial reconnaissance that the collision with the enemy was likely to take place on the line Cambridge-Saffron Walden, and his aim now shifted to driving the enemy not south-eastwards as originally intended but north-westwards toward the 'impassable area' of fenland north of Cambridge. He would change his line of supply to Ipswich so that he could turn the enemy right flank. An intercepted message revealed that Grierson was aware, by means of aircraft reconnaissance, of the location of the Red forces, so Haig decided to halt for the night where he was so that his future direction would remain unrevealed to Blue. Further reconnaissance by BETA in the afternoon confirmed that Blue defenders were digging trenches at Cherry Hinton, south-east of Cambridge. Meanwhile, the Red air force suffered its first setback when one aeroplane, No. 209, came down too low, was fired on and was deemed by the umpires to have been destroyed, ending its

4 Notes on British and French Manoeuvres 1912, prepared by Canadian Officers in accordance with the instructions of the Hon, The Minister of Militia and Defence.

5 The ensuing narrative is largely based on the official Report on Army Manoeuvres 1912, TNA WO 279/47.

6 A. Whitmarsh, 'British Army Manoeuvres and the Development of Military Aviation, 1910-1913', *War in History,* p.333.

BE2 flown in the 1912 manoeuvres – the pilot is Geoffrey de Havilland. (Alamy)

Maurice Farman 'Longhorn' biplane. (Alamy)

role in the manoeuvres. This was an early warning that for all its advantages, aerial reconnaissance would come at a cost.

At 6:00 a.m., Grierson had sent three forces each containing cavalry, cyclists and motor cyclists across the Great Eastern Railway from Bishop's Stortford to Cambridge to locate the enemy's main lines of advance, and by combining this information with reports from his aeroplanes, he was confident by that evening of the location of the

Red forces to his east. Major-General Samuel Lomax's 1st Division was bivouacked around Mildenhall while Major-General Henry Lawson's 2nd Division was around Culford Park, with Allenby and the Cavalry Division to the south of Newmarket. While Grierson's cavalry and cyclists probed the Red defences, the Territorials were ordered to deny Cambridge to the enemy at all costs, using entrenchments and a cyclist company to hold the crossings over the Old West River north of the city. With his infantry now rested and ready to advance, Grierson resolved to go on the offensive the next day.

## Day 2: Wednesday 17 September

The Red army continued its advance on the second day, with Lomax reaching Cowlinge by noon and Lawson's 2nd Division making the first contact with the enemy when Haking's 5th Brigade pushed back Blue mounted units (the Scots Greys) after a short, sharp clash and secured the ridge east of Great Thurlow. Haking was convinced, incorrectly, that he had been confronted by an entire cavalry brigade, and this was the first of a series of failures to identify Blue dispositions which were to cause Haig increasing problems. Fog made observation flights difficult that day, and although aerial reconnaissance was able to confirm the presence of a large force at Great and Little Abington (Rawlinson's 3rd Division) the Blue 4th Division (Snow) was able to evade all Red attempts to spot it. This was despite Red having been handed a stroke of luck when a Blue cyclist had been captured by a troop of the 3rd Hussars carrying a copy of Blue 3rd Division orders. He swiftly tore up the paper, but it was pieced together and sent to Red GHQ. When Haig received the captured orders at 7:00 p.m. he found that they stated that 'the 4th Division is about Saffron Walden' but in the absence of confirmation by aerial reconnaissance he chose to ignore this, perhaps suspecting disinformation on the part of the Blue army, even though intelligence from 2nd Division and the cavalry both confirmed that everything else in the document was genuine.

Showing an awareness of the use of camouflage before the word was known in this context, Grierson had issued detailed instructions beforehand on the use of trees and hedges to conceal troops, guns and horses. He recorded with satisfaction in his diary that Snow's division 'lay concealed all day dodging the aeroplanes, and never were seen.'[7] Artillery used their waterproof sheets and branches to hide the guns, and supply wagons were parked 19 yards apart in fields rather than massed together in the open. Edmonds, Snow's Chief of Staff, recorded that whenever a plane came over when they were on the march, the troops either took cover or changed over from the left side of the road to the right, and as a result of this simple ruse 4th Division was on at least one occasion reported as marching in the opposite direction to their actual

7 D. Macdiarmid, *The Life of Lieut. General Sir James Moncrieff Grierson*, p.246.

Cameron Highlanders in Streetly End during the 1912 Manoeuvres.
(© Hildersham History Recorders)

destination. Tents were of an inconspicuous colour and, if Edmonds is to be believed, one officer painted his with blotches to make it look like a cow.[8]

By the end of the day the 3rd Division was in position near Rivey Hill, where it was engaged in desultory fighting all day, and the Territorials remained at Cambridge. The Blue cavalry had been employed aggressively, establishing contact with enemy columns and in one case (the Hampshire Carabiniers) attacking a battalion of Red infantry in flank and rear, capturing a company and severely mauling an artillery battery, thereby confounding expectations of what the Yeomanry were capable of. By contrast the Red cavalry contributed little that day, but was transferred to Haig's left flank, as he was determined to pivot on his right and launch a strong offensive movement with his left. While 1st Division pinned the enemy and 2nd Division closed with them, the cavalry's role would be to complete the offensive by enveloping the enemy's right in the traditional Napoleonic manner. Shaped by his Staff College experience, Haig saw battle in terms of a structured series of phases, and was seeking to employ the cavalry as he would persist in doing later on the Western Front to achieve the final phase, a decisive breakthrough. It should be pointed out that there was nothing intrinsically wrong with the concept of the structured battle, and Haig can hardly be censured for taking the opportunity presented by commanding such a large body of troops for the first time to test his Staff College training.

8 J. Edmonds, *Memoirs* chapter XXII p.12-13.

The 11th Hussars on the road from Streetly End to West Wickham.
(© Hildersham History Recorders)

Unfortunately for Haig, Grierson was not content to stay on the defensive. Faced with the choice of awaiting an attack in a prepared position and hoping to win by a decisive counter-stroke or advancing to attack an enemy which was on the move, he took the latter course and seized the initiative. He was aware that his situation was not without risk, because if Haig knew, as he surely must do, that the Blue left and right wings were widely separated, the Red army could use a holding force to contain one wing while it concentrated against the other in decisive numbers. Grierson intended to use the Territorials as a reserve, but the difficulty of getting such a scattered force together for a night march made it best to delay the move from Cambridge until daylight on the 18th, thereby running the risk that they would not reach the battlefield until it was too late. In the event, Grierson's plan was successful, because the Red failure to locate the Blue cavalry and Snow's 4th Division would have serious consequences for Haig on the final day.

King George V had arrived with Queen Mary early on 17 September by an overnight train from Balmoral. Large and enthusiastic crowds were there to greet his arrival at Trinity College for breakfast with the Master, Dr Butler. A Redcoat was placed in a sentry box in front of the main gates of the college, while the Porter's Lodge was transformed into the Guard Room. While staying at Trinity, the King found time to visit his late brother Albert Victor's old rooms on Neville's Court.[9]

9 M. Calthorpe, *Royal Cambridge*, p.79.

King George V at the 1912 Manoeuuvres at Little Abington, 17 September 1912. (Saanich Archives)

Dressed in the uniform of a Field-Marshal, he was driven out to the manoeuvres before spending most of the day on horseback. He went first to Grierson's headquarters near Linton, rode up and down the lines and picnicked in a hay-field, before being driven across country to the Red headquarters and seeing some action near Haverhill. That night the Vice-Chancellor of the University, R.F. Scott hosted a dinner in the hall of St John's in honour of the foreign attachés.

The next day the King was driven out to Babraham where he met Seely, who has left a colourful account of what he remembered as 'the most uncomfortable day I ever spent; one which was full of minor disasters... a series of misadventures'.[10] According to Seely's version, his horse went lame and the King was kept waiting while a replacement was sent for from the Remount Depot. When it did arrive and they set off to see the representatives of the Dominions, the King suddenly cried out 'I wish you would stop your horse eating my foot.' The King suffered severe bruising and Seely severe embarrassment, which was amplified when they arrived to meet the Ministers of Defence of South Africa and Canada only to find them 'engaged in a fierce boxing match', with some of the rest trying to separate them, and others looking on; General Sam Hughes of Canada landed a powerful blow on the ear of General Christiaan Beyers of South Africa, shouting 'Traitor!' at the top of his voice at the same time.

10 J. Seely, *Fear and Be Slain: Adventures by Land, Sea and Air*, pp.79-81.

While Seely rode between the two men, Beyers shouted 'Let me get at him.' Seely separated the two men and told them to prepare to receive their monarch. The King had seen the whole thing and asked Seely to find out what it was all about. Supposedly, Beyers had claimed that one South African was better than twenty Britons in a fight, to which Hughes had responded that one Canadian was better than twenty South Africans.

The two pugilists had been adversaries on a previous occasion. As a Lieutenant-Colonel, Hughes had raised a battalion during the South African War, in which Beyers had been one of the Boer commanders. After having his commission revoked and being sent home by British commanders who accused him of 'lacking military discipline', the Canadian had pursued a bizarre campaign against the War Office, demanding to be awarded the Victoria Cross for gallantry in South Africa. Beyers, meanwhile, was one of the most intractable of the Afrikaner leaders in the decade after the end of the Boer War, and was widely distrusted by British officers in the South African command (he was not, in fact, the Defence Minister as Seely described him, but Commandant General of the South African Citizen Force). Though not out of character for Hughes, a notoriously combustible and erratic individual, the story of the boxing match seems inherently improbable, and certainly his most recent biographer makes no mention of the matter.[11]

Beyers certainly made no reference to any such incident in a letter of 23 September to his superior, Defence Minister Jan Smuts, instead giving the impression that he had been received well both at the British manoeuvres and at those of the Swiss and German armies, which he had also attended: 'I had a pleasant talk with the Kaiser, also with King George V.' Beyers informed Smuts that he had attended the dinner for attachés and foreign officers at the end of the Cambridgeshire manoeuvres: 'I spoke on this occasion. French, Seely and others liked my speech. I mention this because no reporters were present.'[12] When war broke out in August 1914, Beyers would lead the pro-German rebellion against the South African government; it is invariably stated that his decision to do so was influenced by his calculation that the Germans would win a quick victory against the British, and that this was due to the positive impression he had gained of the German army at their 1912 manoeuvres.[13] In the event the rebellion was swiftly crushed by South African Prime Minister Louis Botha and Beyers was drowned in the Vaal River trying to escape, while Hughes would continue to be a thorn in the side of the British and Canadian governments until his dismissal in 1916. When Seely published his account in 1931, both Beyers and Hughes were dead and unable to provide their version of events, and Beyers as a traitor was a convenient target for an entertaining story. If nothing else, the episode is a useful reminder of the

11 R. Haycock, *Sam Hughes: The Public Career of a Controversial Canadian, 1885-1916.*
12 W.K. Hancock and Jean van der Poel (eds), *Selections from the Smuts Papers Volume III*, pp.105-106.
13 A. Cruise, *Louis Botha's War*, p.14.

fact that there was an important Dominion perspective to Britain's preparations for war.

Along with Hughes, seven Canadian officers were present at the manoeuvres, a larger contingent than any other nation. They were part of a sizeable group of Militia officers Hughes had brought over on a two month trip which took in visits to the School of Musketry at Hythe, Portsmouth Harbour and the Woolwich Arsenal and the Vickers and Maxim works as well as dividing the party so Hughes and two of the officers could visit the French manoeuvres in Touraine, which took place in the second week of September. The trip was much criticised back in Canada on grounds of cost, not least because of the presence of the officers' wives.[14] An extremely detailed report was compiled as a result of this visit, with the biggest impression on the Militia officers being made, unsurprisingly, by the performance of the Territorial detachment: 'being a selected brigade of Infantry from the territorial force, both officers and men appeared to realise that they were on trial alongside the regular troops, and they created a good impression among the Canadian officers from the way they carried out the task entrusted to them.'[15] They also expressed admiration for the march discipline and resilience of the infantry. Figures compiled by the DMT would later reveal that a total of 120 men fell out on the march with sore or blistered feet at some point during the three days. This figure was made up as follows: 1st Division – 14, 2nd Division – 47, 3rd Division – 11, 4th Division – 26, Territorials – 22. Most men had just completed divisional manoeuvres and so were used to hard marching. The numbers falling out on the retreat from Mons two years later would be significantly higher.

The King's presence on the 17th brought with it large crowds of spectators, though *The Times* regretted that there was little to see on a grey, misty day with low clouds – 'for the public seeking spectacular effect the culminating period of the approach march of the two armies was perhaps dull.'[16] As in 1904, enormous excitement was caused in a part of the country unused to seeing so many troops. *The Cambridge Daily News* coined a new term to describe the ferment of interest – manoeuvreitis. 'If there is such a complaint, I have got it… … for the past fortnight, nay month, it's been nothing but manoeuvre this and manoeuvre that.'[17] Children in Fulbourn were given the day off school so they could go and see the action. Haverhill came to a standstill when the King visited it, with the town's leading employer, Gurteen's textiles factory, closed for the occasion, and there is a commemorative stone at All Saints' church, Horseheath which marks the spot where he had lunch on the following day.

Large crowds can certainly be seen in a remarkable piece of film footage which affords us a unique view of the 1912 manoeuvres in progress. A four-minute newsreel, *The Warwick Bioscope Chronicle*, recorded some of the events, with the King's arrival to

14 A. Capon, *His Faults Lie Gently The Incredible Sam Hughes*, p.47.
15 Notes on British and French Manoeuvres 1912, prepared by Canadian Officers.
16 *The Times*, 18 September 1912.
17 *The Cambridge Daily News*, 13 September 1912.

King George V talking to General Foch during the 1912 manoeuvres. Grierson and Seely are standing behind the King.

observe the Blue army in action unsurprisingly chosen as the item which would excite most interest among the public. King George's arrival by car, with a group of army officers in attendance, including French and Robertson, is followed by a sequence in which the King is shown riding past the cameraman Ager Junior with mounted officers escorting him. This probably took place at the Abington crossroads at Linton. One gets a clear sense of what a difficult task of crowd management the civil and military police faced, and why the army complained about the way the public got in the way of their activities. The film also shows a number of elements of the manoeuvres which are of considerable interest. Numerous foreign attachés are shown, using binoculars and telescopes to follow the action, while a British soldier can be seen signalling with a heliograph. Following this, a 12 pounder field gun positioned in a clover field fires artillery rounds – at close range judging by the angle of the gun's barrel – and is enveloped in a cloud of smoke. Infantry are then shown 'finding cover and firing from trenches'. Soldiers run forward into an open field and throw themselves to the ground, before beginning to dig a bank of cover from which to conduct their defence. This is a graphic demonstration of just how shallow the trenches they dug in 1912 were, and just how different from the ones we are used to seeing in images from the Western Front.[18]

18 Warwick Bioscope Chronicle: His Majesty's Manoeuvres, East Anglian Film Archive, http://www.eafa.org.uk/catalogue/543 Accessed 30/8/2015.

**Day 3: Thursday 18 September**

The final day of the Manoeuvres began with dense mist in the valleys of the Cam and the Granta. This affected Blue plans, Grierson deciding to postpone his proposed movement eastwards, but by 8:00 a.m. the weather had improved sufficiently for orders to be issued to move as soon as possible. With the fog thicker than the previous day, the amount of information which could be gleaned from aerial reconnaissance was limited. By now Haig could see that the Blue orders captured the previous day had been genuine, as Rawlinson's division was still where the orders said they would be, and he was desperate to discover what 4th Division was up to. The captured orders had said that Snow's two brigades would bivouac at Saffron Walden but aerial reconnaissance from Aeroplane 201 at 6:00 p.m. on the 17th had revealed that 'no troops were visible at Saffron Walden'. Aeroplane 207 reported to Red GHQ that at 7:00 a.m. on the 18th the country east of Saffron Walden as far as Hempstead was clear of the enemy, although 'smouldering fires were seen two miles to the east and north-east of Saffron Walden.' The first sighting of the Blue 4th Division was not made until 8:45 a.m. when Aeroplane 403 reported one enemy infantry brigade with an artillery battery halted under cover of a wood to the east of Saffron Walden. This report did not reach GHQ until 10:00 a.m., suggesting that even if the new technology offered great opportunities, that did not necessarily mean speedier identification of the enemy, at

'Telegraphy in a Turnip Field', Horseheath, September 1912. (Saanich Archives)

least for the high command. More effective was the Blue airship GAMMA, which was equipped with wireless telegraphy, enabling almost immediate communication, though Grierson lamented the fact that the airship could not receive wireless messages, so that communication was one way.

Haig's orders for Allenby were delivered in person by Captain Charteris of the GHQ Staff, who had been told by Haig to lay particular stress on the need to locate 4th Division. The Red cavalry moved off in search of Snow's division unaware that they were already within a few hundred yards of the Blue advance guards. Edmonds later remarked that Allenby had 'wandered off into the blue, having brushed past the 4th Division without noticing it'.[19] Having been ordered to move west to locate 4th Division, Allenby soon began to receive frantic messages indicating the presence of enemy cavalry to his front: Colonel W.T. Willcox of the 3rd Hussars reported that 'he is being attacked by cyclists and mounted troops in increasing numbers and has been compelled to retire', while the 16th Lancers reported 'one regiment hostile cavalry 1 mile east of Saffron Walden and a brigade marching on Ashdon'.

Allenby decided that having been ordered to locate Snow's division, he should not get drawn into a duel with the Blue mounted troops. Since the terrain to the west was difficult and occupied by the enemy, it would be easier to move to the south. That way, he could place himself on the enemy's outer flank for his primary role of envelopment of the Blue right wing. When his advance encountered strong resistance from Blue cyclists, Allenby edged south and east, further and further away from the battle. By 6:00 p.m. the Red cavalry had reached Radwinter, where Allenby halted in order to reorganise his brigades and allow them to rest, water and feed. When heavy firing was heard to the north at 3:35 p.m. he decided to move towards the sound of the guns, but his advance was slowed by enemy resistance and the difficult terrain. An attack was about to be launched on the rear of Blue artillery when operations ceased at 5:10 p.m. The Red cavalry had certainly not distinguished itself, having failed to affect in any way the battle which had raged for six hours. Allenby's task had not been helped by persistent communication problems – the Division had not taken its wireless equipment with it and five attempts by GHQ to reach Allenby using despatch riders failed to get through the Blue lines.

It is instructive at this point to examine Haig's *Cavalry Studies* of 1907, which were based on his exercises in India, and which addressed some of the issues with which Allenby struggled five years later in East Anglia. The first study concerned Haig's staff ride of 22-27 February 1904, featuring an invasion across the Jhelum River in the Punjab. The cavalry of the invading Southern Army was facing a weaker enemy cavalry force, but one whose location was uncertain. Haig's orders to the 1st Cavalry Division concluded with the following reminder:

19 Edmonds, op. cit, p.12.

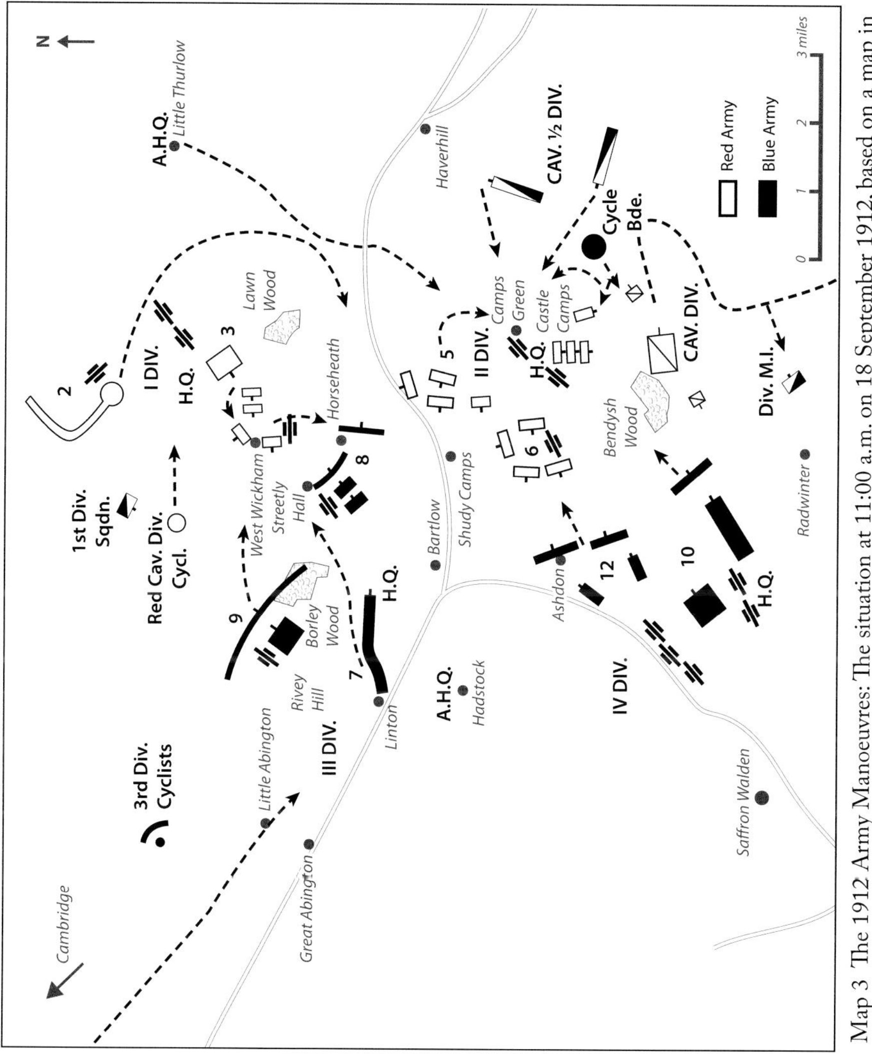


Map 3 The 1912 Army Manoeuvres: The situation at 11:00 a.m. on 18 September 1912, based on a map in the War Office report on the 1912 Army Manoeuvres, WO 27/47.

> In addition to the above specific instructions, it has one other duty for which no instructions are necessary, viz. having once obtained contact with the enemy, never to lose it, and to defeat the main body of the enemy's cavalry, should the latter come within striking distance.[20]

Haig went on to underline the need for the cavalry commander to engage and defeat the enemy cavalry – 'until he does, he cannot carry out the other duties expected of him.'[21] In his study of the Attock ride (March 1906), he restated his belief that the defeat of the enemy's cavalry was the first priority for the cavalry commander, above reconnaissance and above covering the main army's march.[22] In September 1912, Allenby signally failed to carry out Haig's intentions as expressed in his *Cavalry Studies*.

Haig must have been bitterly disappointed by the inability of Allenby and the cavalry to deliver the decisive blow he craved. Numerous explanations have been advanced for the frosty relationship between the two men during the Great War: one of these is Haig's supposed jealousy of the more popular Allenby, dating back to their Staff College days, when Allenby had been elected as Master of the Drag Hounds in preference to Haig. Another involves Allenby's crippling sense of professional insecurity, manifested in extreme shyness. The poor performance of Allenby's Third Army at Gommecourt on 1 July 1916 has also been suggested, though by this stage the tone of their relationship was already firmly in place. The role Allenby played in Haig's discomfiture in September 1912 can plausibly be added to the mix of reasons why these two able but inarticulate men found it hard to communicate with one another. According to Charteris, 'D.H. hardly ever finishes a sentence, and Allenby's sentences, although finished, do not really convey exactly what he means.' The most the two men could manage when together was 'a formal, frigid politeness.'[23]

After his faltering performance at Arras in April 1917, when Third Army proved unable to exploit initial successes, Allenby was removed from command, though this had less to do with Haig than with a loss of confidence in Allenby from his divisional commanders. Ultimately, Allenby's transfer to Palestine enabled him to redeem his reputation and more. Freed from the constraints of the Western Front, he was able to win a spectacular victory by employing cavalry in a series of outflanking manoeuvres of the sort envisaged in Haig's *Cavalry Studies* and planned for in September 1912. Of course, the flat coastal plains of northern Palestine bore little resemblance to the countryside of south Cambridgeshire.

20 Major-General Douglas Haig, *Cavalry Studies: Strategical and Tactical,* p.50. I am grateful to John Hussey for bringing the connection between these studies and the 1912 manoeuvres to my attention.
21 Ibid., p.61.
22 Ibid., p.272.
23 M. Hughes, Edmund Allenby, in I.F.W. Beckett and S.J. Corvi, *Haig's Generals,* pp.20-21.

Whereas Allenby failed to inflict the decisive blow allocated to him, Grierson used his cavalry to pursue more limited but attainable objectives. The mixed bag of Regulars, Yeomanry, cyclists and Mounted Infantry under Briggs' command exceeded all expectations. Briggs handled his brigade with daring and decisiveness, and on several occasions the Blue cavalry demonstrated considerable tactical sophistication. The arguments over British cavalry tactical doctrine had resolved themselves by 1912 into a compromise between shock and firepower, and Briggs' cavalry achieved a model of the instructions contained in *Cavalry Training 1912*. This had stated that 'Cavalry must… keep in close touch with the other arms and take advantage of their progress, offering them as much help as it can with its guns and rifles, and supplementing their decisive attack by sweeping the battlefield with its squadrons.'[24] Having successfully employed a charge the previous day, on 18 September the Blue Cavalry relied on fire from a dismounted position. At Hempstead the Blue cyclists were used in combination with the Scots Greys and Household Cavalry employing rifle and machine gun fire at 1,500 yards range. Later in the day the Scots Greys, along with the Yeomanry of the 1st South West Mounted Brigade, all advancing dismounted, succeeded in capturing the village of Camps Green. The Blue Cavalry also carried out effective reconnaissance. The contrast with the performance of the Red Cavalry could hardly have been more marked.

Haig arrived at 1st Division HQ at 11:15 a.m. and instructed Lomax to hold his position so that Sir Douglas 'might take the offensive with Major-General Lawson and the Cavalry Division on the left'. The most intense fighting of the manoeuvres took place when Maxse's 1st Brigade, supported by eight batteries of artillery, attacked the village of Horseheath and met fierce resistance from Blue's 8th Brigade under Brigadier-General Beauchamp Doran, part of Rawlinson's 3rd Division. One of the officers in Maxse's Brigade recorded in his diary that 'we took part in a terrific battle in + around Horseheath where we should have been slain to a man if there had been real bullets.'[25] Two years later to the day, his battalion would indeed be facing 'real bullets' in the Battle of the Aisne, as would most of the troops engaged at Horseheath. The local press struck the same note: 'Guns by the score were in action on the crest of high ground above, thundering out their streams of shrapnel, and had it been real warfare, Horseheath village would have presented a scene of carnage terrible to contemplate.'[26]

The intervention of Brigadier-General Morland's 2nd Brigade enabled Maxse to capture the village at 1:30 p.m. and push the Blue defenders to the west. The umpires then intervened with a highly contentious judgement, ordering that Horseheath should not be occupied by Maxse's troops so that the Blues would have forty minutes to get clear. Before the time had elapsed, a Blue counter-attack had recaptured Horseheath. This decision, by the umpire attached to the Blue 3rd Division, was

24 S. Badsey, *Doctrine and Reform in the British Cavalry 1880-1918*, p.240.
25 Diary of Capt. J.H. Brocklehurst, Coldstream Guards.
26 *The South West Suffolk Echo*, 21 September 1912.

Artillery firing during the Army Manoeuvres, September 1912. (Saanich Archives)

disputed by the Red umpires and criticised in the Manoeuvre Report. It underlined the difficulty of achieving realism in 'mimic warfare' and the impact of umpiring decisions. French was critical of both divisional commanders for being dragged into a fight for Horseheath, a 'shell trap' overlooked from the higher ground all around and hardly worth the effort expended on its attack and defence. He observed that in the heat of battle, such features sometimes took on an importance regardless of their intrinsic value. Grierson had made the same point as early as 1903, criticising his Commander in Chief, Sir Evelyn Wood, for being too concerned with winning ground as an end in itself, rather than as a means of destroying the enemy.[27] Haig has been attacked for a similar failure to appreciate that 'ground is important only if it enables the killing business to be more efficient.'[28]

At this point in the developing battle, GHQ informed Lomax that Lawson was heavily engaged to the south, and so 1st Division was ordered 'to attack vigorously towards Bartlow' to the south-west. A confused action ensued, during which Rawlinson's position came under severe pressure. It was at this critical juncture that Grierson was able to launch a decisive attack at the crucial point, throwing in his reserves – the Territorials, commanded by Major-General Sir Walter Lindsay,

27 M. Connelly, 'Lieutenant-General Sir James Grierson' in S. Jones, *Stemming the Tide* p.136.
28 S. Bidwell & D. Graham, *Firepower: British Army Weapons and Theories of War 1904-1945*, p.71.

who would briefly command 50th (Northumbrian) Division during the Great War. Sir Percy Radcliffe, later DMT but then a junior staff officer, was sent to meet the Territorials as they arrived from Cambridge. Judging that there was insufficient time to reach the left of 3rd Division as intended, he persuaded Lindsay to commit them on the right of 8th Brigade instead, 'but if I had not known that General Grierson would have backed me I would not have taken the responsibility.'[29]

In approving of Radcliffe's change of direction, 'which he knew was not according to my orders but was appropriate to the situation as it developed',[30] Grierson seems to have absorbed the comments of the Elgin Commission after the Boer War, when commanders had been criticised for being unwilling to allow their subordinates to demonstrate individual initiative.[31] French, in his post-manoeuvre report, judged that Grierson 'wisely, and to a great extent necessarily, appears to have given his divisional commanders a very free hand.'[32] In sharp contrast, Haig was later criticised in the official report for giving Lomax instructions which gave him 'little scope for the exercise of initiative.' Lindsay's advance was delayed half an hour by the weight of civilian motor traffic, and it was not until 4:30 p.m. that two battalions of the Liverpool Regiment had deployed and 20 minutes later that the Territorials' artillery commenced supporting fire. Coming at the same time as an advance by Rawlinson's 7th and 9th Brigades, the intervention of the Territorials transformed the situation on the north of the battlefield.

To the south, Red's 2nd Division began the day on high ground around Castle Camps, totally unaware of the impending attack by Snow's 4th Division. Units of Cavan's 4th (Guards) Brigade ran straight into the advance guard of 11th Brigade, a nasty shock since the Red cavalry, which had just moved off from this spot, had given no warning of the presence of Blue troops. In an impressive demonstration of the potential of an unheralded arm of the service, the crucial intervention in this sector came from Blue cyclists. Their commander, Colonel Lawford placed a battalion in the firing line to pour enfilade fire at a range of 300 yards into the thick lines of Cavan's Guardsmen while he secured his right with another battalion and artillery. The result was inevitable and decisive: completely outflanked, Cavan's Brigade was forced to withdraw as was 6th Brigade on its right, commanded by 'New Zealand' Davies. Major Pereira of the 2nd Coldstream Guards, part of Cavan's Brigade, recorded in his diary that 'we were on the left of the position and were erratically marched in all directions.'[33]

By now Haig had arrived at Lawson's HQ and recognised the seriousness of the situation, since 'the enemy came up with such rapidity from the south against General

29 Macdiarmid, op. cit., p.246.
30 TNA WO 279/47, Report on Army Manoeuvres 1912, p.172.
31 S. Jones, *From Boer War to World War: Tactical Reform of the British Army, 1902-1914*, p.43.
32 TNA WO279/47, Report on Army Manoeuvres 1912, p.56.
33 Diary of Major C. Pereira, Coldstream Guards Archive, Wellington Barracks.

Lawson, that not only was any idea of a counter-stroke out of the question, but it was doubtful whether General Lawson would not have to fall back.' Still pinning his hopes on a decisive blow from Allenby's cavalry, he ordered Lawson to hold on and cabled Lomax to counterattack towards Bartlow. As the afternoon wore on, the position on the Red left became increasingly critical – Cavan was fighting on three sides of a square against the Blue 11th Brigade while Davies was under attack from 12th Brigade, and Haking's 5th Brigade had its hands full with the Blue Cavalry. Snow still had 10th Brigade, which had not fired a shot all day, in reserve when the fighting stopped, a day ahead of schedule, a decision French justified by the two sides being hopelessly entangled. Edmonds would later claim in his unpublished memoirs that 'Haig was so completely outmanoeuvred that the operations were brought to a premature end,'[34] while the press argued that the halt to proceedings was due to the impact of new technology: 'Critics attribute the premature end of the manoeuvres to the perfection of aerial scouting, the efficiency of the wireless service, and aeroplanes nullifying tactical movements.'[35]

When cease fire was called at 5:10 p.m. the Blue Army clearly held the upper hand. 1st Division's right flank had been driven back and the attack against the right of 3rd Division had been held. 4th Division, ably assisted by the Blue cavalry, had met with success against 2nd Division, and Snow still had 10th Brigade intact to deal with the Red cavalry in its rear. The Territorials were just starting to make their presence felt. The Secretary of State for War was in no doubt as to the outcome: 'Grierson had so completely achieved his purpose in the scheme which had been set that there was nothing left to fight about.'[36] Sir Ian Hamilton, writing to Repington of *The Times* a fortnight later, was in agreement: 'Red should have been told that whilst the issue of the actual battle was still uncertain, he had been so severely handled (that) any advance on London would be inconceivable.'[37]

It must be conceded that Grierson's troops, as the defenders, did enjoy a significant advantage when it came to aerial reconnaissance because Haig's troops were attacking and had to march greater distances in three divisional columns, so were much easier to spot from the air, but Grierson had defeated a stronger and better organised enemy which possessed the moral advantage of being the invader, while 'the appearance of the concealed division was the sensation of the manoeuvres.' However, it had been a close-run thing; Grierson had run the risk of his two infantry divisions being defeated in detail and the Territorials had only just arrived in time. It was with some relief that the victorious commander recorded in his diary that night:

34 R. Holmes, *The Little Field-Marshal: Sir John French*, p.149.
35 *The Nelson Colonist*, 21 September 1912
36 J. Seely, *Fear and Be Slain: Adventures by Land, Sea and Air*, p.81.
37 A.J.A. Morris, *Reporting the First World War: Charles Repington, The Times and the Great War*, p.161.

> My orders were that the enemy should be attacked wherever met… This brought on a desperate battle in which at the end the situation was much as opposite, and I having the 10th Brigade still in hand, and having thrown the Territorials in at the right moment got the best of it. The Red Cavalry was somewhere in my rear, but it didn't matter for the battle was gained… Bivouacked with cavalry camp at Linton and stood champagne in honour of our victory (such as it was!)[38]

The post-manoeuvres conference took place the next day in the Hall of Trinity College in the presence of King George V and a hundred senior officers. Haig, an Oxford man, ambiguously recorded the conference as having taken place in Trinity Hall, a different college altogether.[39] A notoriously poor public speaker, Haig spoke first, and he confided to his diary that 'I think my remarks were well received'.[40] The truth seems to have been quite otherwise. His Assistant Military Secretary John Charteris recounted the tale of Haig's woeful effort:

> Although Haig had written out a clear and convincing statement of his views, and held the paper in his hand throughout the Conference, he did not refer to it when he spoke, but to the dismay of his staff attempted to extemporise. In the effort he became totally unintelligible and unbearably dull. The University dignitaries soon fell fast asleep. Haig's friends became more and more uncomfortable; only he himself seemed totally unconscious of his failure.[41]

Haig's terse self-confidence conflicts sharply with what others, taking their cue from Charteris, have seen as his disastrous performance at the conference (though it is worth pointing out that Repington, no friend of Haig, made no mention of it in *The Times*, instead stating that Haig had given 'a full and clear account of the operations of the Red army').[42] If Charteris is to be believed, it is likely that Haig's notes had been written on the assumption of an evenly matched battle, and when this turned out not to be the case he attempted to improvise, with dire consequences. The editor of Haig's diaries for this period, his grandson Douglas Scott, has speculated that Charteris may not have actually been present at the conference, since he was in Bulgaria by 2 October and may have had to set out for his destination before 19 September – 'perhaps his version of events was received second hand.'[43] However, there seems no reason to believe that Charteris could not have attended the conference in Cambridge on 19 September and still have been in Sofia 13 days later. The military correspondent

38 Macdiarmid, op. cit., p. 247
39 Douglas Haig (ed. Douglas Scott), *The Preparatory Prologue: Diaries and Letters 1861-1914*, p.319.
40 Ibid.
41 Brig-Gen. J. Charteris, *Field-Marshal Earl Haig*, pp. 65-66.
42 *The Times*, 21 September 1912.
43 Haig, op. cit., p.320.

for *The Daily Telegraph*, Ellis Ashmead-Bartlett, rushing like Charteris to the Balkans to be there in time for the outbreak of the First Balkan War, left Charing Cross on the evening of 2 October and was in Constantinople five days later, having travelled via Vienna and Bucharest.[44]

In any event, Haig's performance in the manoeuvres does not appear to have counted against him in the long run, any more than did his stumbling display at the conference – certainly George V, himself proverbially inarticulate, did not hold it against him.[45] Grierson followed Haig, and according to one eye-witness, 'he looked like victory'. His speech was full of the touches of humour which endeared him to his men – describing his orders to the 4th Division to lie concealed about Saffron Walden, 'I told them to look as like toadstools as they could and to make noises like oysters.'[46] Edmonds remembered Grierson's comments slightly differently, recording in his memoirs that when the King asked him how the troops had remained hidden, Grierson replied 'Your Majesty, they put on their waterproof sheets over their heads and made noises like mushrooms.'[47] He professed himself delighted to hear from Haig that the Blue attack 'upset his calculations, took the initiative from him and compelled him to conform to his opponent's movements.'[48]

Sir John French summed up and the King expressed his pleasure at spending three days among his troops: 'The present system of training was conducted on sound lines, and he especially observed the intense keenness and earnestness of purpose shown by the Army.'[49] The Royal Party then departed for Balmoral and the Conference ended, leaving the press to reflect on what they had witnessed. Grierson, who once told Tom Bridges that the medals on his ample chest commemorated many battles fought with knife and fork, headed to London to dine at the Metropole as the guest of the Army Council at a dinner given to the foreign and dominion officers who had attended the manoeuvres, and the next day he lunched at the French Embassy to meet General Foch, Colonel Reboul and other French officers, a good example of the role manoeuvres played in the increasingly strong links being forged with Britain's ally.

Opinions on what had taken place during the three days of the Army Manoeuvres differed widely. The Secretary of State for War believed that 'This was, in my judgement, the only sham fight which I have ever seen that had any relation to actual warfare,' his presence at the manoeuvres contradicting Edmonds' claim that he had never witnessed a minister at any set of pre-war army manoeuvres.[50] Presumably Edmonds also did not see Winston Churchill at either the 1910 manoeuvres, which he attended when Home Secretary, or those in 1913, by which time he had become

44 E. Ashmead-Bartlett, *With the Turks in Thrace,* p.4.
45 G. Mead, *The Good Soldier: The Biography of Douglas Haig*, p.166.
46 Macdiarmid, op. cit., pp.247-248.
47 Edmonds, op. cit., p.13.
48 TNA WO279/47, Report on Army Manoeuvres 1912, p.175.
49 *The Times,* 20 September 1912.
50 Seely, op. cit., p.81.

First Lord of the Admiralty.[51] Another minister, John Burns at the Local Government Board, was one of many to send Grierson a message of appreciation: 'Hearty congratulations. The 4th Division was splendid. March discipline and walking first rate.'[52]

Seely found another feature of the manoeuvres highly gratifying. Addressing a public meeting at Dumfries the following week, he paid tribute to the good behaviour of the British soldier, recounting that the Chief Constable of Cambridgeshire had reported not one single case of crime from the 48,400 troops who had taken part. Seely went on to make the remarkable claim that 'the great British Army was now not only the bravest but it was the best-behaved army the world had ever seen.' He argued that this fact had ended any reason for the old-fashioned prejudice among the working and middle classes against the Army on the grounds of its conduct: 'the British private of today is in truth a kind of grown-up public-school boy.'[53] Whatever the truth of this statement, it is clear that Seely was well aware of the importance of the British Army's relationship with civilian society at a time when industrial unrest was a potential source of conflict.

However, *The Cambridge Daily News* was unimpressed by the spectacle, employing the headline 'Futility of Manoeuvres. Bullet the Only Real Umpire', and describing fighting 'of a most determined and sanguinary character. Had the troops been fighting "with the gloves off", the carnage must have been appalling.'[54] *The New York Herald* observed, with little originality, that the spectacle was 'magnificent, but it was not war.'[55] *The Times* was more positive, stating that the King 'was able to witness one of the most realistic battles ever produced in England or any other country.' It had been carried out 'in the most realistically scientific manner possible in the circumstances of mimic war', the newspaper explaining that by 'realistic' it did not mean mock heroics, but that contacts between the two armies were developed 'in the most realistically scientific manner possible in the circumstances of mimic war.'[56]

The Liberal press was agreed that Grierson had won a famous victory – he had 'demonstrated his genius', whereas 'with regard to General Haig's strategy, no particular brilliance could be claimed. He worked almost entirely by the training manuals, and his every move could be foreseen and met.'[57] While *The Manchester Guardian* recognised that Haig was 'second to none as a staff officer, or as a man able to handle a unit which he can retain in his own hand', it concluded that Grierson 'seems to exhibit himself as the bigger man, more in the manner of Lord Kitchener, who can impose his will on circumstances, and whose mental ramifications are wider.'[58] *The Times* was more

51 Edmonds, op. cit., p.18.
52 Macdiarmid, op. cit., p.248.
53 *The Spectator*, 3 October 1912.
54 *The Cambridge Daily News*, 19 September 1912.
55 Calthorpe, op. cit., p.80.
56 *The Times*, 19 September 1912.
57 *The Westminster Gazette*, 20 September 1912.
58 *The Manchester Guardian*, 20 September 1912.

equivocal about the outcome, 'no very decisive result having apparently been achieved by either force.' The even-handedness of its correspondent, Charles Repington, is significant given his personal dislike of Haig, whereas Grierson was an old friend, the two men having been colleagues in the Intelligence Department between 1889 and 1894 alongside Henry Wilson, Charles Callwell and Wully Robinson under their patron Brackenbury.[59]

In making this judgement, Repington was sticking closely to French's cautious summing-up at the conference, in which he appeared at pains to be even-handed. Manoeuvre reports tended to be marked by a reluctance to indulge in overt criticism of individuals, especially those in high command. Even so, French was critical of Haig's failure to seize two opportunities. The first of these was at the outset when the Red force could have moved on Cambridge and achieved an early success against the Territorials or the Blue cavalry, though Repington in *The Times* felt that given the fact that the 1st Division would have had to march thirty miles to reach Cambridge, it would have been a gamble to attack entrenched troops with tired men. 'There are not many commanders who will take these risks, and we must give up placing Scotsmen at the head of our armies if we wish such risks to be run.'[60] Haig recorded that French's 'criticisms especially on the strategic value of Cambridge were not much thought of'.[61] Haig's second missed opportunity was on the final day when there was a chance to drive into the six mile gap which had developed between Grierson's two infantry divisions, something for which French took Grierson to task. He felt that the Blue commander had run a considerable risk and had only got away with it because of 'the failure of the Red cavalry' and his own bold decision to advance to meet the enemy.

French also argued that by becoming involved in the local situation when he went to visit Lawson, Haig possibly lost sight of the wider battle and failed to see that he still had a chance of winning in the north (he and Lawson also only narrowly avoided being captured by Blue cavalry). This would be a recurrent problem for commanders on the Western Front, caught between a desire to get close to the action and the need to stay back out of harm's way and to remain in control of communications. French was of the opinion that at Horseheath, once Haig had given the orders for 1st Division to attack at 1:30 p.m. 'his control over the battle practically ceased.'[62]

Ironically, French would himself be guilty of a similar fault at Loos in September 1915, severely limiting his ability to influence that battle at the crucial moment. Besides the issue of command and control, there were considerable dangers inherent in generals being too far forward; Lomax, commanding 1st Division as he had in 1912,

59 See W. Ryan, *Lieutenant Colonel Charles a Court Repington: A Study in the Interaction of Personality, the Press, and Power*. For Brackenbury, see C. Brice, *The Thinking Man's Soldier: The Life and Career of Sir Henry Brackenbury 1837-1914.*

60 *The Times*, 7 October 1912.

61 Haig, op. cit., p.319.

62 WO 279/47, Report on Army Manoeuvres 1912, p.56.

Major-General Lomax and the staff of 1st Division at Eriswell, September 1912. (Saanich Archives)

would be mortally wounded at Ypres on 31 October 1914.[63] The same shell which claimed his life when it hit 1st Division headquarters at Hooge Chateau killed five staff officers and fatally wounded another. The commander of 2nd Division, Monro, was severely concussed in the same explosion and the command structure of Haig's 1st Corps was temporarily disrupted at a critical moment.[64] When eight generals became casualties (five of them killed, two wounded and one taken prisoner) in the space of nine days at Loos, action was taken to prevent a repeat of these damaging losses: 'These are losses which the Army can ill afford, and the Field Marshal/C in C desires to call attention to the necessity of guarding against a tendency by senior officers such as Corps and Divisional Commanders to take up positions too far forward when fighting is in progress.'[65] French was belatedly heeding one of the lessons of the 1912 manoeuvres, though too late to save his own job – within six weeks he would be replaced by Haig.

63 Lomax died in a London nursing home from the effects of his wounds on 10 April 1915.
64 F. Davies and G. Maddocks, *Bloody Red Tabs, General Officer Casualties of the Great War, 1914-1918,* p.5. The authors provide a comprehensive demolition of the insidious myth of 'chateau generals'.
65 GHQ Order of 3 October 1915, quoted in Davies and Maddocks, op. cit., p.29-30.

Manoeuvre reports can be frustratingly anodyne in their judgements – it could, after all, be embarrassing and damaging to morale to have public criticism of commanders who were known to be the men who would lead their country's troops into battle in the event of war – but one does not have to read too carefully between the lines to detect in French's comments the fact that he had not been impressed with the display from his fellow cavalrymen Haig and Allenby. In particular, the performance of Allenby and his division had been extremely disappointing: 'It is not clear why Major-General Allenby did not locate the 4th Division much earlier than he appears to have done, or why, after he had located it, he did not act more directly against the enemy's flank.'[66]

French's main criticism of Haig was that he had mishandled his attack on Grierson on the last day of the manoeuvres. If Haig's plan was to fix the Blue army with inferior forces (1st Division) and throw a superior force (2nd Division and the cavalry) at the right flank, French argued that he should have engaged the enemy with the 1st Division rather than wait for the enemy to attack first. When the Blue 4th Division was at last detected, Haig should have responded to swiftly changing circumstances by changing his plan, a criticism which would be made on a number of occasions on the Western Front. His implication was that Haig was fixated on his scheme for a flanking movement – 'Even after it had become evident that the 2nd Division was being attacked in force, he still relied on his cavalry to turn the scale.' Even here, much of the blame still rested with Allenby – things could have turned out very differently 'if he had been well served by his cavalry.' [67]

In an attempt to be even-handed in his summing-up which sat uncomfortably with the general tenor of his report, French concluded that it was impossible to decide what the ultimate results of the battle might have been, although he conceded that the advantage clearly lay with the Blue army when fighting ceased.[68] In a letter to Lord Esher on 24 September, French revealed his true opinions of Haig and Grierson. While reiterating that it was impossible to decide which of the two men would have been successful in a real war ('I think Grierson can always make up his mind more quickly than the other, but neither of them is what I should call a "dasher"'), French concluded that 'I should judge that certainly Haig, and perhaps Grierson, will always shine more and show to greater advantage as superior Staff Officers than as Commanders.'[69]

That Grierson's victory was not as clear-cut as has often been claimed is supported by the contemporary observation of a presumably well-informed member of the public, W. Burnard Wilder, Rector of St Mary's, Great Bradley: 'Army Manoeuvres were held in the Eastern Counties. The important battle took place at West Wickham

66 WO 279 / 47, Report on Army Manoeuvres 1912, p.56.
67 Ibid., pp.54-55.
68 Ibid., p.56.
69 Major G. French, *The Life of Field-Marshal Sir John French, First Earl of Ypres*, p.184.

and Horseheath – result indecisive.'[70] The few historians to have examined the 1912 Manoeuvres in any detail have tended to accept uncritically Philip Warner's judgement that Haig was 'completely out-generalled by Grierson'[71], who 'successfully outwitted Haig with some aplomb'.[72] Terraine concedes that 1912 was 'not a shining hour for Haig'[73] and Reid judges that Grierson 'proved infinitely the better general.'[74] However, if this is a valid judgement, the old notion that Haig's defeat was in large measure due to his failure to employ aerial reconnaissance is well wide of the mark and now generally discredited.[75] It is therefore disappointing to read Alan Mallinson's judgement in 2014 that 'there is no doubt that during the manoeuvres Haig underestimated the value of aircraft in observation.'[76] Mallinson does concede that 1912 taught Haig useful lessons on the usefulness of aerial reconnaissance, and he seeks to excuse Haig's supposed lack of familiarity with the new technology by pointing out that he had left Britain to take up his role as Chief of the Indian General Staff before Louis Blériot's cross-channel flight in July 1909 (actually, Haig sailed for Bombay on 8 October 1909) and returned in January 1912, when the Royal Flying Corps had not yet been founded. In fact, Haig had been impressed by the potential of aircraft as early as 1911, commenting positively on their first use in manoeuvres in India – a Bristol Boxkite aeroplane with Captain William Sefton Brancker of the Royal Artillery as observer was able to locate the enemy and report back even though his landing strip was captured while he was in flight. Haig went so far as to suggest that, if the finances were available, officers should be sent back to England to learn to fly, so that an army flying school could be opened in India; an Indian Flying Corps was duly set up in 1914.[77]

The idea that Haig was resistant to new technology such as aeroplanes, machine-guns and, later, tanks, is too tempting to resist for those determined to portray him as a hidebound 'donkey' but is far from the truth.[78] If anything, Haig was far too dependent on reports from aircraft in September 1912, his plan of attack on the final day being predicated on the assumption that the enemy right flank was undefended, when actually the 4th Division was there.[79] In a classic case of cognitive dissonance, where an individual rejects information which conflicts with his expectations of what will happen, excessive faith in aerial reconnaissance led Haig to discount all

70 Notes of W. Burnard Wilder, http://www.great-bradley.suffolk.gov.uk/History/wilderfamily2.htm Accessed 30/8/2015.
71 P. Warner, *Field Marshal Earl Haig*, p.108.
72 Connelly, op. cit., p.141.
73 J. Terraine, *Douglas Haig – The Educated Soldier*, p.53.
74 W. Reid, *Douglas Haig: Architect of Victory*, pp.159-160.
75 For example Warner, op. cit., p.108.
76 Allan Mallinson, *1914: Fight the Good Fight: Britain, the Army and the Coming of the First World War*, p.60.
77 A. Whitmarsh, op. cit., p.333.
78 See J. Laffin, *British Butchers and Bunglers of World War One* and D. Winter, *Haig's Command: A Reassessment* for two egregious examples.
79 G. Sheffield, *Douglas Haig: From the Somme to Victory*, p.63.

evidence to the contrary, such as the captured orders. In this respect he displayed a trait which would be seen from him and other generals on the Western Front, where they had unrealistic expectations of the results that emergent technology, such as gas and tanks, could deliver.[80] As Major C.H. Foulkes, the creator of the Gas Brigade, would later recall: 'Throughout the war I found officers of high rank almost too receptive of novel proposals, especially when they were based on anything mysterious or scientific.'[81] The fact that British generals were keen to embrace new weapons and tactical developments might not sit comfortably with proponents of the 'donkeys' school of thought, but it should not surprise us – why would any commander neglect a development which could hand them an advantage, however small, in such a well-matched struggle?

If there was disagreement about the outcome of the manoeuvres, the one thing on which everyone could agree was that they had proved the importance of the new aerial dimension of warfare. Grierson stated that 'Our first attempt at using aircraft in manoeuvres has, I think, been an unqualified success, and the impression left on my mind is that their use has revolutionised the art of war', reiterating these views at a meeting of the Aeronautical Society later in 1912.[82] It was generally agreed that the aeroplanes had obtained as much information in three hours as would have taken a cavalry division three days to procure.[83]

In fact, such was the impact made by aircraft that there was widespread concern that too much reliance could be placed on their reconnaissance, as Haig had discovered to his cost. He admitted in the official report that 'The information brought in was, as a rule, so reliable that there seems a danger in taking it for granted that if no enemy are seen by the observers none are there.' French, so keen a supporter of aircraft that he served as chairman of the Aeronautical Society before the war, agreed, arguing that 'the experience of the manoeuvres shows that even in favourable conditions it is not safe to rely entirely upon air reconnaissance, and that in close country it is possible for large bodies of troops to escape observation for some time.'[84] This was not resistance to technological change but a sensible response to a new element in warfare which was very much in its early stages.

While air power was still a new factor for commanders to consider, two developments from the 1912 manoeuvres suggest that some were already thinking in surprisingly innovative terms. First, the use of a feint made by the Red cavalry in the hope that the move would be observed by Blue aircraft shows that even at this stage experiments were being made with deception measures against aerial observation, which

80 D. Jordan and G. Sheffield, 'Douglas Haig and Airpower' in P. Gray and S. Cox (eds.), *Airpower Leadership: Theory and Practice*, p.268.
81 Quoted in P. Griffith, *Battle Tactics of the Western Front: The British Army's Art of Attack 1916-1918*, p. 111.
82 Andrew Whitmarsh, op. cit., p.339.
83 Notes on British and French Manoeuvres 1912, prepared by Canadian Officers.
84 TNA WO 279/47, Report on Army Manoeuvres 1912, p.61.

would be a regular feature of the First World War.[85] Second, some perceptive individuals had already seen that the value of aircraft for reconnaissance was such that it would be vital to gain control of the air, either by shooting down enemy aircraft from the ground or arming one's own aircraft. Assistant umpire R.F. Hare declared that 'so long as hostile aircraft are hovering over one's troops all movements are liable to be seen and reported and, therefore, the first step in war will be to get rid of the hostile aircraft.' The side which could do this or 'keeps the last aeroplane afloat' would win. Frederick Sykes, Officer Commanding the Military Wing of the Royal Flying Corps, even spoke of the need to establish 'command of the air', in an early use of that term nine years before it was popularised by the work of Giulio Douhet.[86]

The celebrated aviation pioneer and Wild West showman Samuel Cody, who was present throughout the manoeuvres, told *The Times* that the next step in aircraft design would have to be 'a fighting aeroplane', which he envisaged as a large vessel which would be used 'like a battleship'.[87] This smacked of fantasy to some participants at the 1913 Staff Conference, such as Colonel Du Cane, who argued that a better use of their time would be to concentrate on attaining superiority in numbers, speed and training: 'It seems to require the imagination of a writer of fiction to picture the possibilities of fighting in the air.' French, however, disagreed, summing up the deliberations in favour of exploring the possibility of arming aeroplanes and suggesting that with the rapid pace of change in this emergent technology 'we may see most extraordinary developments', a claim which would be amply borne out between 1914 and 1918.[88]

As well as the broader lessons of the manoeuvres, specific improvements were suggested for the new arm. The need for camouflage had been shown, prompting the statement in the 1912 Memorandum on Army Training that commanders needed to use the forthcoming training season to practise the concealment of troops from airmen. Given the increasing likelihood of fire from the ground, the need for aircraft to bear distinguishing markings, probably along national lines, had also been demonstrated. Another lesson was that such was the strain of flying that both planes and pilots needed more leisure time between sorties. Hare noted that only one of the five officers in the Red squadron looked as if he could go on by the end of the third day, the strain being discernible from the involuntary movement of their hands when they returned from flying. Tired pilots and mechanics 'cannot be expected to notice the frayed wire that means disaster.' The need for trained observers had also been underlined. Sykes concluded that 'untrained observers are quite useless.' Strategic reconnaissance tended to be carried out in favourable conditions for spotting, when troops

85 Whitmarsh, op. cit., pp.336-337.
86 TNA WO 279/47, Report on Army Manoeuvres 1912, pp.170-172.
87 *The Times*, 20 September 1912.
88 TNA WO 279/48, Report of a Conference of General Staff Officers at the Royal Military College 1913, p.57.

were in column and, in dry weather, throwing up dust. However, tactical reconnaissance depended on spotting troops after they had left the roads, and so was much harder to do. Troops were easily distinguishable when there were few roads and railways, and in these conditions it was easiest to allow the pilot to find his way and leave the observer to observe, but when over difficult country, with small winding roads, as in southern Cambridgeshire, a specific objective was necessary and the observer needed to direct the pilot. A pilot observer in a single aeroplane might gather useful information if there were large strategic movements over easy country, but otherwise a two-man crew was essential.[89]

Finally, thought had to be given to guarding landing places, an especial problem in a fast-moving campaign and one which the Royal Flying Corps would encounter in August 1914. On the final day a patrol of Scots Greys had the chance to take prisoner some of the Red squadron officers, along with an aeroplane, at their landing place near Little Thurlow, while the Blues lost two of their machines, both captured by a detachment of Hussars when one came down for adjustments and the other landed to render assistance.

*The Cambridge Daily News* found grounds for national pride, comparing the manoeuvres with those of the German Army in the same week, in which over one hundred troops had died, or so the newspaper claimed. There were actually seven fatalities connected to the East Anglia manoeuvres: two civilian drivers working for the ASC were killed in accidents, Thomas Hurlock when a horse lashed out and fractured his skull at Little Thurlow, and Charles Carter when an RAMC transport wagon he was driving overturned and crushed him. The inquest at the Cock Inn, Thurlow, decided that he was responsible for the accident because of his careless driving, although the jurors were keen to place on record the absence of a warning notice at the top of the steep hill.[90]

Four of the fatalities were members of the fledgling RFC, which had been established only five months earlier. Early on 6 September Captain Patrick Hamilton set off from Wallingford in his 100 horsepower two-seater Deperdussin monoplane with Lieutenant Athole Wyness Stuart as observer. Their mission was to engage in reconnaissance for cavalry divisional manoeuvres and land at Welwyn, but disaster struck over the Hertfordshire village of Willian, near Hitchin. When the Gnôme rotary engine failed catastrophically at an altitude of 500 feet, the aeroplane was seen to rock violently and the wings collapsed, causing it to crash, killing both men instantly.[91] Mr Walter Brett, landlord of the George and Dragon in Graveley, would later tell the Coroner: 'I saw the aircraft wobbling about. It dipped and then came a report like a gun. Then the aircraft seemed to collapse altogether. I was too horrified to look any more … I ran down and found the officers lying with the machine on top of

89 TNA WO 279/47, Report on Army Manoeuvres 1912, p.73.
90 *The South West Suffolk Echo*, 28 September 1912, 21 September 1912.
91 *Flight Magazine*, 14 September 1912.

The funeral of Lieutenant Wyness Stuart and Captain Hamilton at St. Saviour's Church, Hitchin. (Saanich Archives)

them.' Their funeral took place with full military honours amidst large crowds at St Saviour's Church in Hitchin, after which the coffins were carried on gun carriages to the railway station, from where Captain Hamilton's remains were sent to Hythe in Kent and those of Lieutenant Wyness Stuart to Mells in Somerset.[92]

Four days later, pilot Lieutenant Charles Bettington and observer Lieutenant Edward Hotchkiss were killed when their Bristol Coanda monoplane broke up and crashed at Wolvercote, just outside Oxford, as they were en route from Larkhill on Salisbury Plain to Hardwick, near Cambridge to take part in the manoeuvres.[93] An inquest held the following day at the Plough Inn, Wolvercote, heard that as the monoplane circled before landing at Port Meadow, a flying wire came loose and tore a hole in the fabric, sending the machine out of control. Bettington was thrown from the monoplane while Hotchkiss went down with it. Thousands turned out for their funeral procession through Oxford on 13 September, with the Royal Artillery supplying a gun-limber for the coffins, accompanied by the band of the 4th Oxfordshire and Buckinghamshire Light Infantry.

92 'Terrible air fatality', http://www.hertsmemories.org.uk/content/herts-history. Accessed 30/8/2015.

93 *Flight Magazine*, 12 October 1912.

Memorial at Wolvercote to two casualties of the 1912 manoeuvres. (Author's collection)

Both of these crashes were marked by memorials near the spot where they took place. In November 1912, a small granite obelisk bearing the names of Hamilton and Wyness Stuart was erected about half a mile away from the Hertfordshire crash site by the side of the road from Willian to Great Wymondley. Hamilton's mother laid a wreath of chrysanthemums upon the obelisk and the uniforms of the dead officers were buried under it. Their Flight Commander, Major Robert Brooke-Popham, who had accompanied Hamilton's aeroplane as far as the Hertfordshire border in his own biplane, made a short speech in which he told his listeners:

> The careless child and the weary wayfarer will pass along this road, look at this stone, read this inscription and realise that they, too, have a duty to perform. They will know that patriotism is not an empty word and that Englishmen are still ready to lay down their lives in the service of their country.

Meanwhile in Oxfordshire, a 'penny collection' sponsored by the *Oxford Journal Illustrated* raised £31.10.11d, enough for a memorial tablet set into the wall near Wolvercote Bridge as well as a plaque in Wolvercote Parish Church. The plaque was made from a sheet of metal taken from the crashed aircraft.[94] At this point there

94 P.F. Wright, *The Royal Flying Corps in Oxfordshire 1912-1918*, pp.8-11.

had been 12 monoplane fatalities in 1912 alone, and the Wolvercote crash led to a five-month War Office ban on monoplanes in favour of biplanes and had two serious consequences: a long-standing prejudice against monoplanes among the authorities and a halt to design progress at a crucial stage of this rapidly developing technology.

The last of the seven fatalities was Private Harry Sebastian, a Territorial from the 5th Battalion, Liverpool Regiment, who drowned while bathing in the Cam. In addition, Gunner T. Smith was blinded in both eyes and had his right arm amputated at Addenbrooke's Hospital after the breech of his gun blew up.[95] More prosaically, Major-General Rawlinson also had to visit Addenbrooke's, in his case following a fall from his horse; after an X-ray revealed a dislocated toe, he was hastening back to take command of his division within an hour.[96] There was also one civilian casualty of the manoeuvres, a boy named Alfred Anderson, who was accidentally shot in the leg, though not seriously hurt, when he got in the way of soldiers who were firing from behind some hedges at Fulbourn.[97] Manoeuvres may not have involved real fighting, but that is not to say that they were entirely bloodless.

95 *The Cambridge Daily News*, 19 September 1912.
96 *The Cambridge Daily News*, 17 September 1912.
97 *Fulbourn Chronicle 1901-1930*, compiled by D. Crane (1983), 27 September 1912.

5

# The 1913 Exercise

*The Northampton Mercury* could hardly have been accused of underselling the story when it informed its readers of the forthcoming 1913 Army Manoeuvres:

> Two armies will march across North Bucks – horse, foot, artillery and all the latest equipment in scouting and commissariat. Motor traction will be utilised as never before by the British Army. Military cyclists will be present in great numbers. Wireless telegraphy will be installed. Most fascinating of all will be the flying craft, which are expected to operate on a much greater scale than has ever before been possible.[1]

The September 1913 manoeuvres, held in Buckinghamshire and South Northamptonshire, were of a very different nature to those of the previous year; although the number of troops involved was roughly the same (45,000 men), this was not a contest between two evenly matched forces led by men of equal rank and stature as it had been in 1912. Instead an army of 38,322 men commanded by the Chief of the Imperial General Staff, Sir John French, took on a much smaller force of 7,075 men under the command of Major-General Charles Monro which was there primarily 'to provide a target'.[2] It is interesting to note that the Army Council opted in 1913 for such an exercise given that in their 1909 Memorandum on Army Training they had pointed out the shortcomings of operations against a skeleton enemy and had stated their intention of providing opportunities in future for generals to learn how to cope with encounter battles.[3] Such an exercise was also out of step with contemporary developments on the continent, for example von Moltke's attempts in Germany

1 *The Northampton Mercury*, 5 September 1913.
2 TNA WO 279/52, Report on Army Manoeuvres 1913, p.5. On Monro, see P. Crowley, *Loyal to Empire: The Life of Sir Charles Monro 1860–1929.*
3 TNA WO 27/508, Annual Report of the Inspector General of the Home Forces for 1912, p.8.

to return decision-making to the commanders and to avoid an overly prescriptive approach to manoeuvres.[4]

The 1913 exercise should be seen as a strategic development of 1912, but whereas 1912 had had as its avowed purpose a test of leadership for the rival generals and their staffs, 1913 was more specific in its aims and circumscribed in its scope. Troops would manoeuvre in a pre-arranged fashion following a specific timetable and leading to a pre-ordained outcome, an assault on an entrenched enemy. Significantly, the War Office chose the term 'Exercise' for the events of 21–25 September 1913 rather than 'Manoeuvres' as 1912 had been described, reflecting the different nature of what was being attempted. The War Office clearly set out the Exercise's aims as testing the following:

1. The working of General Headquarters and Army Headquarters at war establishments.
2. An approach march of one cavalry division and two armies, each army consisting of two divisions and marching on one road.
3. The supply of the above force, the supply services being confined to the same roads as those on which the troops marched.
4. Cavalry covering the advance of the same force (Brown) against a slightly superior White cavalry reinforced by cyclists, necessitating the support of the Brown cavalry by the advanced guards of the main columns.
5. The driving-in of the enemy's advanced troops.
6. The deployment of the Brown Force for attack.
7. An attack on an enemy in an entrenched position.
8. Organisation of a pursuit.
9. Breaking off the pursuit, and a sudden change of direction.[5]

The relevance of a number of these objects to the conditions of warfare the same generals and troops would face eleven months later is striking, not least the organisation of a pursuit (number 8) and an attack on entrenched troops (number 7). By definition the organisation of a pursuit entailed practice of a retreat for the other side, and this was something a number of leading figures had been calling for in recent years. This suggests a High Command that was able to adjust its training to respond to changing circumstances, even though, according to Snow, General Douglas, Inspector General of Forces, did not like troops to be exercised in anything to do with a retreat or retirement as he thought that it was bad for morale.[6] Monro, who led the White force carrying out the retreat in 1913, would find himself within a year commanding a

4 E. Dorn Brose, *The Kaiser's Army: The Politics of Military Technology in Germany during the Machine Age, 1870-1918*, p.154.
5 TNA WO 279/52, Report on Army Manoeuvres 1913, p.5.
6 TNA CAB 45/129, General T. Snow, Account of the Retreat of 1914, p.3.

division engaged in the retreat from Mons. In his report on his last staff tour in India in 1911 before he returned to take up the Aldershot command, Douglas Haig had warned of the need for Britain to be prepared for unforeseen circumstances in any future war, as 'no plan of operations can with any safety include more than the first collision with the enemy.' Contrasting the British Army's doctrinal flexibility with the German General Staff's commitment to the doctrine of envelopment, he concluded, 'An army trained to march long distances, to manoeuvre quickly, and to fight with the utmost determination will be a suitable instrument in the hands of a competent commander.'[7] On his return to Britain, he insisted on training I Corps at Aldershot in the 'manoeuvre in retreat', which Duff Cooper claimed 'proved of vital importance to the officers and men who were compelled to take part in the inevitable retreat from Mons.'[8]

Haig was not alone in this prescience. Lieutenant-Colonel W.H. Greenly gave a lecture to 2nd Division in January 1913 on 'Employment of Cavalry in a Retreat', and would have the opportunity to draw upon his thinking in his role as GSO1 of 2nd Cavalry Division in 1914.[9] Allenby had repeatedly warned of the need to practise retreats and Brigadier-General Haldane had visited Spain in April 1911 in order to follow the route of Moore's retreat from Corunna, covering every yard by foot and bicycle.[10] He wrote after the war that 'Though our training in peace time had largely consisted of practice in offensive operations, the question of retreats… had always interested me.'[11] Snow, who was Haldane's divisional commander, records that the 1913 divisional manoeuvres included a scheme which involved Snow retreating hastily for 24 hours, pursued by Rawlinson's 3rd Division. They got into such difficulties that Snow decided that further practice was needed, so the next staff ride in the spring of 1914 included a scheme for a 2½ day retirement following an unsuccessful action: 'We learned a great deal and found out what the difficulties were which we should encounter in a retreat, how to overcome those difficulties, and the duties of every member of the staff in a retreat.'[12]

It should, of course, be borne in mind that there may be an element of hindsight in what Haldane and Snow have to say in their post-war recollections. More significantly, Robertson had used an address to the students at the Staff College in November 1912 to make the same point: 'Our regulations justly lay stress on the value of the offensive; but think what may be the effect of this teaching upon the troops if it alone is given, when they are ordered to retire instead of going forward – that is, to abandon that method of war by which alone, according to the training they have previously received,

7 J. Charteris, *Field-Marshal Earl Haig*, pp.55-56.
8 D. Cooper, *Haig*, vol. 2, p.436.
9 G. Sheffield, 'Lieutenant-General Sir Douglas Haig' in S. Jones (ed.), *Stemming the Tide*, p.118.
10 B. Gardner, *Allenby*, pp.63, 70.
11 Lt. Gen. Sir A. Haldane, *A Brigade of the Old Army 1914*, pp.67-68.
12 General T. Snow, edited by Dan Snow and Mark Pottle, *The Confusion of Command: The War Memoirs of Lieutenant General Sir Thomas D'Oyly Snow, 1914-1915*, p.7.

decisive victory can be achieved.'[13] After Robertson became DMT at the War Office in the autumn of 1913, he was in a position to remedy the lack of practice at retreating, and for the 1914 manoeuvres he designed a scheme whereby a force would retire in the face of a more numerous enemy. This scheme included the crossing of the river Severn by the retreating force, an extremely difficult operation for a retreating force hard-pressed by its pursuers.

By the time these manoeuvres were meant to take place, September 1914, the BEF had already taken part in the retreat from Mons, and Robertson's performance as QMG, in particular his resourceful approach to supply problems, suggested that he, for one, had given a good deal of thought beforehand to the problems facing a retreating army: 'The study of the manoeuvres we had planned was most helpful to me during the first few weeks of the war, when I was hard put to it to keep the troops supplied with what they needed.'[14] Edmonds credited a staff exercise devoted to the study of a retirement set by Robertson in March 1914 with giving the staff of 4th Division the confidence to carry out the procedure after Le Cateau five months later: 'We had come to the conclusion that a force can always break off an action and retire ... provided that it had sufficient notice to enable the train, the ammunition columns, etc. to be sent off in good time, the roads clear and under good traffic control.'[15]

The 'General Idea' for the 1913 exercise posited three imaginary states. Greenland (which would not actually feature in the exercise) in the northwest was the traditional enemy of Whiteland in the centre. Brownland in the south-west had been on friendly terms with Whiteland but tension had developed due to growing commercial and military rivalry. Greenland and Brownland had taken the opportunity of a sudden quarrel to mobilise for a simultaneous offensive against the Whites. Observant readers could have been forgiven for imagining that Green could be identified with France, Brown with Britain and Whiteland with Germany, while the German military attaché Martin Renner, who would attend the manoeuvres, may have spotted a strong resemblance to the Schlieffen Plan, which had dominated German war planning and strategy since 1905 in an attempt to strike first against one enemy in order to avoid a war on two fronts as Whiteland now faced.

According to the 'General Idea', Greenland and Brownland declared war on Whiteland on 20 September. Whiteland had 12 infantry divisions and 3 cavalry divisions, Green 9 infantry divisions and 4 cavalry divisions, Brown 4 infantry divisions and 1 cavalry division. In reality, Greenland and the White forces facing it were ignored, and the exercise was restricted to the clash between Brown and White forces.

French's Brown force consisted of two armies, the First, under Haig, being made up of the two divisions of his Aldershot command under Lomax and Lawson (essentially the same force that Haig had commanded the previous year) and the Second,

13 Haldane, op. cit., p. 68.
14 Field-Marshal Sir W. Robertson, *From Private to Field-Marshal*, p.195.
15 J. Edmonds, *Memoirs*, chapter XXII p.15.

under Paget, comprising Rawlinson's (3rd) and Snow's (4th) divisions. Allenby once again commanded the Cavalry Division, made up of three brigades. The 1st would be commanded by Briggs, who had performed so well against Allenby the previous year, the 2nd by de Lisle and the 4th by Bingham. Grierson, who had been earmarked for the position of French's Chief of Staff in the event of war, would get his chance to show what he could do in this role.

Monro's White force contained two skeleton infantry divisions which existed only on paper and also included cavalry, horse artillery, two battalions of cyclists and detachments of the Royal Engineers, Army Service Corps and RAMC. His cavalry division was made up of two brigades of Yeomanry and a regular brigade containing his best troops, the Scots Greys, 19th Hussars and a composite Household Regiment (the Life Guards and Blues). *The Northampton Mercury* expected great things of Monro's White army, 'including the most splendid cavalry in the British Army and imposing squadrons of aircraft. How can this defensive force, brilliantly generalled, confound the schemes of the invaders?'[16]

The aerial component was considerably expanded from the previous year; where the 1912 manoeuvres had employed fifteen aircraft and two airships, in 1913 the Brown and White Armies between them had 33 aeroplanes, half the total that would cross to France with the BEF in August 1914. The White force had two airships attached to it, DELTA and ETA, as well as the preponderance of the available aircraft, as befitted a defensive force faced with the task of scouting for a larger enemy on the move. Monro could call upon the seven aeroplanes of No. 4 Squadron of the RFC (two BE2a, one Breguet and four Maurice Farman Longhorns) and the eight of No. 5 Squadron (four BE2a and four Maurice Farman) as well as six aircraft from the Naval Wing (three Short biplanes, a Sopwith, a Caudron and a Blériot monoplane). French's Brown army was supplied with twelve aeroplanes of No. 3 Squadron (four Blériot XI, three Henri Farman F. 20, 3 BE2a, a BE3, and a BE4). All were supplied with foreign engines, Gnôme, Renault or Astro-Daimler. An unprecedented 67 officers and 490 men of the RFC would take part, the latter figure including large numbers of ground crew, one of whom, Air Mechanic James McCudden, would become one of the greatest 'aces' of the Great War. *The Northampton Mercury* confidently averred that 'those who still think the British Army have no aviation equipment will have their eyes opened during the week of the Army Exercises.'[17]

The commander of No. 3 Squadron, Brooke-Popham, made extensive preparations for the exercise, overseeing the painting of the undersides of the wings of the planes to enable identification from the ground (this was before the introduction of roundels) and supervising the provision of transport and tentage for men and aircraft servicing. He personally reconnoitred the area of operations to identify landing fields for the

16 *The Northampton Mercury*, 5 September 1913
17 *The Northampton Mercury*, The Army Manoeuvres and the Royal Visit (September 1913) p.16.

squadron, drawing on the connections made in his time with the Ox and Bucks LI to drop in for dinner at the homes of local landowners before returning to his base at Netheravon on Salisbury Plain each night in the taxi hired by the army for the purpose.[18]

Once again, the exercise took place in September as the culmination of the training period, which included the usual inter-divisional exercises. According to Edmonds, not necessarily the most reliable of sources, it was during these exercises that Rawlinson was 'as usual, caught cheating', gaining information on the whereabouts of Snow's 4th division through underhand means, employing cyclists wearing umpires' armbands as well as sending out Lady Rawlinson in a car. Finding a large concentration of officers in Wincanton, she reported back to her husband that the town was occupied by Snow and his division. Rawlinson duly launched a night attack against the town, only to discover that it was in fact French and his staff that had been spotted. The Field Marshal was not amused to be awoken in the middle of the night. Snow and Rawlinson had been contemporaries at the Staff College, and both were apparently 'foxes' who were not above such a ruse in order to steal a march on a rival.[19] Whether or not the story is true, and it does appear to stretch credulity, it does give a flavour of the intense competition between commanders keen to make an impact in a small army where opportunities for promotion at the top level were relatively few.

The first signs of the exciting events about to unfold were detected as early as the start of August, when the Royal Engineers began to arrive to prepare camp-sites around Wolverton, which would serve as the railhead for 2nd Army's advance. Towards the end of the month, they were joined by men from the Army Service Corps, the RFA and the Royal Irish Fusiliers. The first troop train arrived with little fuss at Wolverton Railway station on 29 August at 3:00 a.m., bearing 200 men of the 2nd Lancashire Fusiliers, but by the time the fourth train had pulled into the platform, which had been specially extended for the occasion, and disembarked the Seaforth Highlanders, accompanied by their pipes, people were up and about and hundreds of women and children had turned out to see the excitement, despite thick fog. This included the ladies employed at Messrs. McCorquodale's Printing Works, who turned out en masse.[20] According to the local paper, the Highlanders were the troops that most locals wanted to see:

> Wolverton practically lost its identity on Friday night, and looked more like a garrison town on the occasion of a Bank Holiday crush. Swarms of people flocked into the town from the neighbourhood, and the scene was a very animated one, and it was early evident, as the people began to mass along the thoroughfares,

18 F. Hanford, *First Landing: The story of the first arrival of aircraft on the Halton Estate*, p.2.
19 J. Edmonds, *Memoirs*, chapter XXII p.13.
20 J. Sunderland and M. Webb, *All the Business of War: the British Army Exercises of 1913, the British Expeditionary Force and the Great War*, p.14.

> that the Gordon Highlanders were in for a great reception. Their train was scheduled to arrive at 8.35, but it was after nine o'clock before they marched out of the goods yard. The people could be counted in thousands, and the press was tremendous. The pipers chose a merry Highland tune for their march to the Stony Stratford camp. Many people present, owing to the immense crowd, were able to see but little beyond the glengarries of the men, the majority of whom wore their traditional kilts. On entering the Stratford Road there were scenes of unparalleled enthusiasm. Cheer upon cheer rent the air as they stepped briskly, along the road to Stratford, which was thickly lined with people for about half a mile… It seemed as if there could be nothing to damp the enthusiasm of the crowd. A fine rain was falling, and the conditions underfoot were miserable; but the people cared little. For a long time after the Highlanders had disappeared the streets bore the appearance of holiday making. Soldiers in squads walked about the town singing snatches of songs and rag-times.[21]

On the first day 576 men and 590 horses, 38 wagons and 46 guns passed though Stony Stratford. The next day the 3rd Royal Dublin Fusiliers arrived at the station, as well as the 2nd Lancashire Fusiliers, the 1st Royal Warwicks, the 1st Berkshires, the East Lancashires, the East Durhams, the Gordon Highlanders, the 1st Rifle Brigade, the South Wales Borderers, the 1st King's Own and the 2nd Royal Dublins. Soon, 13,000 troops had passed through the station and were camped between Stony Stratford and Wolverton. This impressive logistical feat had been achieved by Mr Lancelot Horne, Assistant Superintendent of the Line for the LNWR, assisted by Mr Brinklow, stationmaster at Wolverton and a team of superintendents and inspectors called in from as far afield as Crewe, Euston and Rugby. Horne would later serve as acting secretary to the Railway Executive Committee and played a major part in the remarkably smooth process of mobilisation in August 1914:

> No hitch whatever occurred so far as the London and North-Western Company was concerned in carrying out, not only the pre-arranged programme, but also the additional movements which have been arranged at short notice. The time-keeping of the trains has been excellent both from a traffic department and loco-motive department standpoint, and the entraining and detraining at the various stations on the London and North-Western line were successfully carried out in every case.[22]

The troops were soon busy with daytime exercises, including constructing pontoon bridges across the Ouse, and with night marches. Brigade training began on Monday 1 September. *The Wolverton Express* described 12th Brigade's training in some detail.

21 *The Wolverton Express*, 5 September 1913.
22 L.W. Horne, quoted in G. Darroch, *Deeds of a Great Railway* p.27.

An attack in close country was carried out as a drill. The Brigade deployed, advanced across country in preparatory formation, and attacked the fleeing defenders. With heavy rain continuing all day, three minor casualties were suffered. One was a private of the South Wales Borderers, who slipped while running on the wet grass and dislocated his knee. An artilleryman was thrown off the limber through the gun skidding, whilst a rider, whose horse slipped, sprained his ankle through the fall. Around 3:00 p.m. operations ceased and the men, equipped with waterproof capes, made their way back to camp. By now the ground had turned into a quagmire, but the weather had improved by Wednesday 3 September.

As well as brigade and divisional training, there was also time for socialising with the local people, for whom the arrival of so many soldiers was the most exciting thing to have happened in their locality for many a year. On Sunday the camps were thrown open to the public, the one at Warren Hill, near Stony Stratford being the most popular attraction as it contained the Highland Regiments. A Grand Military Tournament at Wolverton Park was attended by more than 6,000 people. Troops were treated to a round of dances, church choir concerts and smoking concerts, not to mention visits to public houses, which were understandably keen to attract their custom, while officers attended coffee mornings or dinner parties at one or other of the area's many country houses. Sir John French already knew the area well, having had the tenancy 30 years before of Yew Tree Farm on the Hanslope estate; he now chose as his headquarters Hanslope Park, home of Mrs Edward Watts, his wife's recently widowed sister.

One local landowner who was especially hospitable was Alfred de Rothschild, who hired marquees and caterers in order to treat the men of the Black Watch, the Munster Fusiliers and the Grenadier Guards who stayed on his land on the Halton estate near Aylesbury; Halton was renowned for the lavishness of its house parties. The authorities had to dissuade Rothschild from his original offer to provide all meals for the troops but were unable to prevent him from laying on a high tea for three evenings running. 3,000 men of the 1st Infantry Brigade were treated to pies, cold meats, bread and butter, along with tea, beer and mineral water. The officers were, unsurprisingly, treated even more lavishly, with ample supplies of champagne available in their mess marquee, which was decorated with floral arrangements.[23]

By 21 September divisional training had ended and attention shifted to the main event. French's Brown forces began to cross the frontier with Whiteland at 9:00 a.m., with 1st Army bivouacked on the Uxbridge to Aylesbury road, 2nd Army between Watford and Leighton Buzzard and the cavalry between Woburn and Thame. The intelligence provided to Brown was that the bulk of the White forces were engaged in fighting the Greens, but two White divisions (the ones which only existed on paper) were advancing from Nuneaton and Coventry towards Daventry and cyclists had been spotted moving south-east from Rugby. The White cavalry was moving towards Towcester while their cyclists were ordered to hold onto the crossings on the Ouse.

23 Hanford, op. cit., p.4.

The Royal Flying Corps attached to the White Army had a crucial role to play in spotting the Brown infantry columns and observing the Brown cavalry.

## Day 1: Monday 22 September

The Brown Army advanced north-west, Haig's 1st Army towards Winslow and Paget's 2nd Army towards Bletchley, while Allenby and the cavalry division protected the left flank. The River Ouse was the site selected for the first clash, a clear sign that this was a much more stage-managed and artificial exercise than 1912 which had essentially been allowed to develop without interference from the Director. The purpose of this opening encounter was to give the Brown forces practice in crossing a river while under fire. The 2nd and 4th cavalry brigades advanced towards the River Ouse at 9:00 a.m. from Mursley and Swanbourne in Buckinghamshire and by midday they were ready to begin the assault. Two squadrons of 4th Dragoon Guards were ordered to secure the bridge at Thornton to enable the river to be crossed. They reached the bridge only to find it 'destroyed', but then found a drawbridge which the Whites had failed to destroy and had crossed the Ouse by 12:15 p.m., to be joined by the rest of de Lisle's 2nd brigade two hours later. 1st Brigade, under Briggs, moved up to Thornton to cover the river crossing, in company with a section of the RHA. White cyclists from Norfolk and Essex spotted the advance upon Thornton Hall and slowed the advance down by blowing up a bridge across the carriage drive. There was a heated exchange of fire, with the cavalry dismounting and lying on the tennis courts and lawn in order to drive back the defenders, a textbook example of the use of cavalry in a dismounted role. Both brigades were now ordered to attack Buckingham at 3:45 p.m., the 1st Brigade from the south and the 2nd from the east. There was little opposition and the town was occupied by 4:30 p.m., with White forces withdrawing northward towards Daventry. With only a small force of cavalry and cyclists, there was little they could do against overwhelming odds.

White aircraft performed well in a reconnaissance role on the first day, although they could not take off before 10:30 a.m. due to mist. Once up in the air they detected an enemy column four miles long advancing towards Fenny Stratford – this was Paget's 2nd Army. Meanwhile, the White airship, DELTA, took off at 7:15 a.m. and went as far south as Marlow, detecting two enemy columns heading north. Near Great Missenden she came under 'attack' from a Brown Blériot monoplane and a B.E. biplane which succeeded in intercepting the airship within six minutes of its being spotted, but the umpires ruled DELTA's crew to have driven off the two aeroplanes circling it because the airship 'was a steadier platform', a judgement which would not be borne out in the coming war.[24] At this stage neither airships nor aeroplanes had armaments, though Grierson had not been alone in suggesting after the 1912

24 *The Times*, 4 October 1913.

manoeuvres that they should be considering this option. This was an early warning of the shape of things to come in the next few years and of the vulnerability of airships in a reconnaissance role.

By the end of the first day, French knew that there were two White skeleton divisions close to Daventry and that their cavalry was at Towcester. For the next day he ordered a general advance, with the 2nd Army moving on Towcester via Stony Stratford and the 1st Army on Helmdon via Buckingham. The cavalry would cover the advance.

## Day 2: Tuesday 23 September

The second day would see a great deal more fighting. It would also witness the sole recorded death associated with the 1913 manoeuvres; once again, it involved a fatal accident to a driver in the ASC. Nineteen-year old Arthur Deyton was involved in an accident near Boddington on the afternoon of 23 September and died in Northampton General Hospital the same evening. Corporal F. Dean of the 16th Company ASC testified to the inquest that the horse that Deyton was riding plunged and he fell on its neck, at which the horse reared and fell back on him.[25]

The second day of the manoeuvres was marked by the arrival of King George V and Queen Mary, who visited Northampton, where they unveiled a statue of Edward VII and were presented with a loyal address from the borough, as well as inspecting troops in the market square. They were then taken to Towcester to see the White Army's defence of the town. The inhabitants had never seen anything like it – Towcester School was 'closed owing to the Army Manoeuvres' as their daybook recorded. The children were excited by the pageantry and, no doubt, by the day off lessons.

Towcester was occupied by the White cavalry, which had bivouacked in 800 white bell tents erected on Easton Neston Park just south of the town. There were about 5,000 troops in all, men of the London and Midland Mounted Brigades, the 6th and 8th cyclists, two battalions of the RHA, a detachment of the ASC, a field ambulance and a transport column. Present to witness what was expected to be the week's decisive clash were Prince Alexander of Teck (Queen Mary's brother), Sir John French, Colonel Seely and Winston Churchill. All were staying at Park House, overlooking Towcester Racecourse, while the King and Queen were the guests of Earl Spencer at Althorp House.

By 10:00 a.m. Allenby was aware that the 1st and 3rd Divisions had secured Buckingham and Stony Stratford. He was also aware that enemy cavalry were around Silverstone, with two battalions of White infantry reported at Whittlebury. He ordered the 1st Brigade, with a section from the RHA, to support his 2nd Brigade, which was reported to be engaged with White cavalry. As they advanced they were

25 *The Northampton Mercury*, 26 September 1913.

fired on by enemy cavalry but drove them back thanks to artillery support. Twenty minutes later the 2nd Brigade was relieved by Bingham's 4th Brigade which had been ordered up on its right. The Whites then retreated to Silverstone, which was attacked by all three Brown brigades. By 1:15 p.m. it had been captured along with some prisoners. The White cavalry continued its retreat towards Abthorpe while the Brown cavalry retired to Brackley where it bivouacked.

Haig's 1st Army saw little action on the second day. The 1st Division set off at 6:50 a.m. and the 2nd twenty minutes later; they reached Syresham by 9:00 a.m. after capturing some White Yeomanry and cyclists. They then headed towards Helmdon, bivouacking for the night by 4:00 p.m. and waiting for their supply columns to reach them. To the east, Rawlinson's 3rd Division advanced on Towcester along Watling Street from their encampment at Bletchley, setting off at 7:15 a.m. White cavalry brought the advance to a halt half a mile south of Potterspury. Infantry from the Brown vanguard forced the cavalry to withdraw to a position alongside a White battery, a mile east of Paulerspury. The horse artillery took a pounding while the London Mounted Brigade bore the brunt of the fighting. This position was also untenable, and they had to withdraw again through Paulerspury at about 11:30 a.m., reforming in a strong position south-east of Towcester. This time the Brown advance was halted and heavy casualties sustained. The situation was only saved for Brown by the RFA and by a flanking attack to the right by units of the 7th Infantry Brigade under Brigadier-General McCracken. The White position was overrun and they were pursued through Towcester, streaming back to Pattishall to the north. The 3rd Division, having taken the town, now bivouacked there, with its front line stretching from Stoke Park to Abthorpe. Meanwhile, the 4th Division under Snow had moved up to Stony Stratford and Wolverton.

Monro had already decided that his best chance of success was to delay the Brown advance on the 23rd and the 24th before falling back to a prepared position further north, where he could engage them. He therefore deployed his cavalry on 23 September to harass the Brown cavalry's flanks while his 11th (skeleton) division moved south at 5:00 a.m. and took up position in order to prepare a defensive entrenchment for the whole force. 11th Division would hold off the enemy as long as possible but was ordered not to put itself at risk of being unable to disengage from the enemy. To the west of the Towcester action, White cavalry advanced through Silverstone, removing an enemy cavalry brigade from the Akeley to Whittlebury road. At about 10:30 a.m. it became obvious to their commander that the superior enemy cavalry advancing towards him posed a considerable threat. He followed his orders and withdrew in an orderly manner towards Abthorpe through Silverstone. The North Midland Mounted Brigade covered this withdrawal and suffered the loss of two squadrons in a confused fight.

The White forces were not harassed and kept 11th Division's position screened during the night. They bivouacked at Canons Ashby. One regiment of the North Midland Mounted Brigade was spread out as an advance guard towards Banbury. They reconnoitred the Brown position at Helmdon from 6:00 p.m. Aircraft were sent out for the same purpose, reporting back that an enemy infantry brigade had been

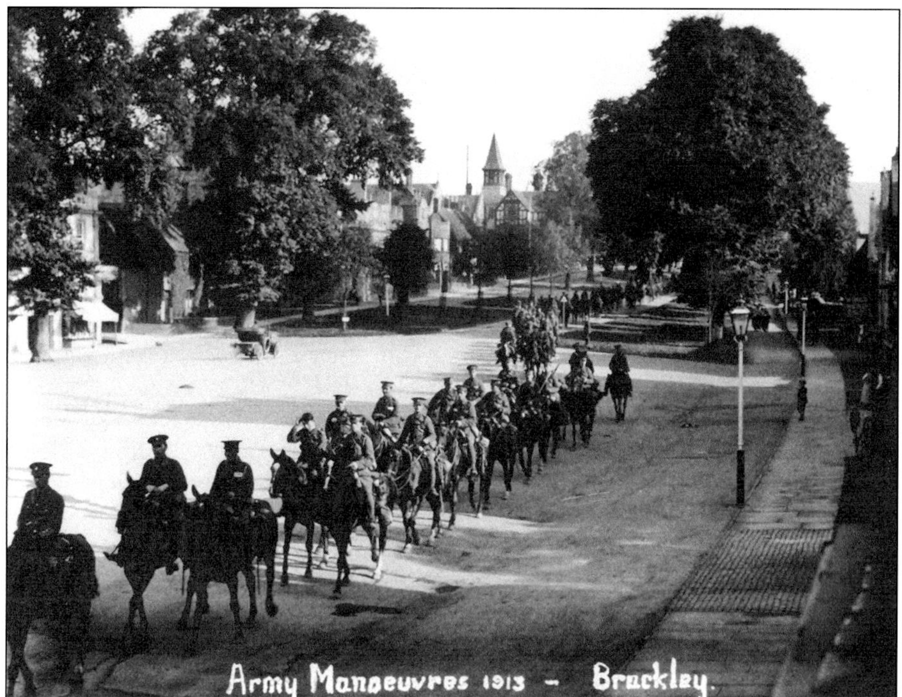


The Scots Greys in Brackley during the 1913 Army Exercise. (Author's collection)

spotted in bivouac near Yardley Gobion with its cavalry near Stowe Park. This led Monro to think that the bulk of the Brown cavalry was on his left, at Stowe Park, rather than at Brackley, posing a threat to his right flank, as was actually the case. Once again, aerial reconnaissance was producing misleading results.

The 24th would witness the resumed Brown advance on Daventry. 2nd Army would leave Towcester at 7:00 a.m. and advance on the right, while 1st Army would be on the left. The cavalry would cover the left flank of the advance and push reconnaissance towards Banbury and Leamington.

**Day 3: Wednesday 24 September**

The White Army was now engaged in a 'fighting retreat', designed to allow an army to withdraw its forces by staged resistance, enabling the main body to retire to a new defensive position. The advance guard of Brown's mounted troops were soon checked by White cyclists at the railway bridge north of Towcester. They were well dug in and could only be removed after the intervention of Brown's Kent cyclists and Brown's leading battalion of infantry. The march could now continue, with the 15th Hussars covering their right flank.

The progress of Rawlinson's 3rd Division towards the River Cherwell was halted by White forces, and the problem was only solved when 8th Brigade under Brigadier-General Doran launched a flanking attack supported by all available Brown artillery. Even then, it required the further support of troops from Snow's 4th Division to force White back. Rawlinson was able to push north while Snow reached Maidford, where his division bivouacked for the night. Rawlinson's division was able to link up with the 1st Division of Haig's 1st Army at Woodford Halse on the River Cherwell, thus closing a dangerous gap which had developed between the two forces. Haig had pushed north, relying on the 1st Battalion Grenadier Guards in the vanguard to defeat a force of White Essex cyclists, before crushing the 30th White Infantry Brigade (a paper unit) with the help of the 2nd Infantry Brigade as its flank guard.

Meanwhile the cavalry had also been heavily engaged. Brown's 1st Brigade had been ordered to make contact with 1st Army around West Farndon, and to do so they sent a squadron to watch the river crossings at Banbury, Cropredy and Wardington while the main force pushed westward, reaching Culworth by 9:00 a.m. There they met fierce opposition from the North Midland Yeomanry, which inflicted heavy losses before withdrawing, leaving the Brown cavalry to camp at Edgcote, site of a battle from the Wars of the Roses (1469). Allenby sent one squadron and a machine gun north of Culworth to establish contact with the 1st Army. By 6:00 p.m. all the elements of French's army had linked up and formed a continuous front facing the White defensive position south of Daventry. Monro suspected that Brown was going to try to turn his right flank using cavalry and 1st Army and this suspicion was confirmed by troop movements spotted by aircraft that evening. Monro had his defensive arrangements in place and at 7:30 p.m. issued his orders for the next day – his troops were to hold their positions on the line Hellidon/Sharman's Hill/Great Everdon/Upper Weedon. 10th Division was notionally on the right and 11th on the left with the London Mounted Brigade at Upper Weedon and the remainder of his cavalry at Priors Marston. If they were forced to retire they would destroy railway and canal bridges and fall back to a position on the line from Braunston to Welton.

### Day 4: Thursday 25 September

Monro occupied a strong position. He had dug trenches and had artillery and cavalry with which to counter-attack, but he was massively outnumbered. Allenby's cavalry was out to the west, in a position to keep an eye on any White reinforcements from the north and west while standing ready to assist Haig's 1st Army in its attack on the White defences. Paget, commanding 2nd Army, had Snow's 4th Division, along with two artillery batteries, which would be held in reserve around Woodford Halse while the three infantry brigades of Rawlinson's 3rd Division were allotted to his frontal attack between Everdon and Fawsley, which began at 8:00 a.m. Within one hour they had come up against strong resistance from well entrenched infantry along with enfiladed artillery fire, in a modest precursor of what would be encountered in the

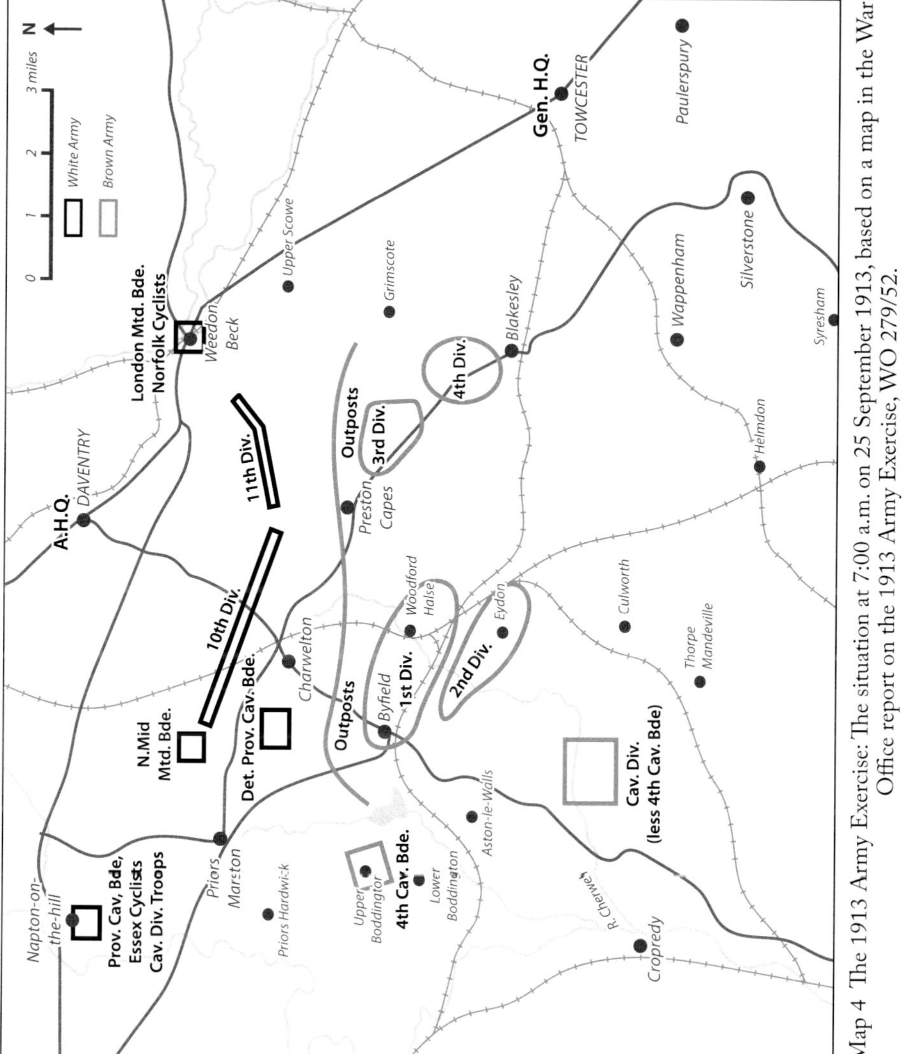


Map 4 The 1913 Army Exercise: The situation at 7:00 a.m. on 25 September 1913, based on a map in the War Office report on the 1913 Army Exercise, WO 279/52.

Great War. 34th Artillery Brigade returned fire and allowed 8th Brigade on the right of 2nd Army to continue the advance. 9th Brigade on Paget's left was held up north of Fawsley by stubborn White resistance. After fierce fighting the Whites were beaten off, and forced back to their trenches on Everdon Hill. Half an hour later they came under attack from 8th Brigade, which took the village following a short but sharp fight and captured many prisoners. Paget was making good progress.

Haig's progress was much slower, partly because he was facing White's best troops, the Scots Greys and Household Cavalry, and partly because he chose to approach his task methodically, first ordering the village of Priors Marston to be taken. This attack was delayed by the need to wait for artillery cover to be in place, and when it was launched Maxse's 1st Brigade, comprised of the Grenadier Guards, the Royal Highlanders and the Munster Fusiliers, was met by fierce fire from the Essex cyclists. The attackers 'took advantage of every possible shelter, but they were exposed to a scathing fire, and, if ball cartridge had been used, many a man would have been stretched dead, before, by sheer weight of numbers, the position was captured.'[26] By 11:10 a.m. 1st and 2nd Brigades, advancing with 3rd Brigade in support, had taken the position and the Essex and Norfolk cyclists had been forced to retreat.

By 1:00 p.m. French must have felt that all was going well. Lawson's 2nd Division was advancing to the west of Cropredy. The Guards Brigade, part of Lomax's 2st Division, had taken Hellidon Hill, and at about 1:30 p.m. Paget prepared for his final assault, ordering up 11th Brigade along with a brigade of RFA in support of 9th Brigade, attacking strongly defended entrenchments around Fawsley. After two hours of fierce resistance, the Brown forces overran the whole White front from Great Everdon to Fawsley. Monro ordered his troops to fall back towards Daventry, improvising a line along the River Nene. Brown's 4th Division was now ordered to carry out the pursuit while 3rd Division re-formed before following. Paget only halted the pursuit when he received news in the late afternoon that two White divisions were detraining at Evesham and Redditch, 40 miles west of Daventry. This was one of those points during manoeuvres when the directing staff intervened to shape events. Concerned that his troops were becoming strung out and vulnerable to attack, Paget halted his army, which bivouacked around Badby.

By this time Haig was at last ready to assault his objective, Sharman's Hill, and ordered 2nd division to attack from the west while six battalions of Lomax's 1st Division advanced on the south and south-west of the hill. Artillery support was provided by three brigades of RFA and a howitzer brigade. The climactic battle was watched by a large crowd of spectators, which included the King, the Duke of Connaught and Lord Roberts, as well as the Queen, who was accompanied by Lady Adelaide Spencer. All were expecting to see Haig deliver a swift *coup de grace*. However, the Brown attackers were surprised by a skilful White counter-attack against their left flank, and a confused action developed. Ultimately, the Whites, facing overwhelming odds, withdrew and

26 *The Northampton Mercury*, 26 September 1913

allowed Haig to take Sharman's Hill at last. To the west, Brown cavalry on 1st Army's flank went into action against White's best troops around 1:00 p.m. The Household Cavalry and Royal Scots Greys attacked the high ground west of Hellidon but were pushed back with heavy losses by Brown cavalry and an RHA battery. The Brown Army went into bivouac at 5:30 p.m. 'no doubt relieved and exalted as well as exhausted.'[27] *The Northampton Mercury* sounded a cautionary note in its assessment of the Brown victory:

> So ended the Battle of Sharman's Hill. Whether it would have ended so had actual war conditions existed only experienced soldiers can say. Certainly it would not have ended so soon. The opinion of many military authorities was that the hill could not have been carried in that style, even with a longer time to do it, without an appalling loss of life.[28]

Twelve months later, Haig's 1st Corps would indeed experience 'an appalling loss of life' when 3,500 casualties, many of them from the same battalions which had taken Sharman's Hill, were suffered in assaulting the German positions on the heights above the River Aisne on 14 September 1914.

## Day 5: Friday 26 September

The decisive action of the manoeuvres had taken place and the Brown forces were now free to exploit their victory. Haig's and Paget's armies were ordered westward from 7:00 a.m. to meet the White divisions detraining at Redditch and Evesham while a cavalry detachment was kept back to monitor the beaten White forces. The Brown 4th Cavalry Brigade was busy scouring all roads leading towards Daventry from the north. All the Whites could manage was some cyclist and cavalry reconnaissance which led to fierce skirmishing in which the Yeomanry once again made a good impression. At 9:00 a.m. it was clear to the umpires that French had achieved his objectives and operations were called to a halt.

The final conference took place in the riding school at Weedon Barracks, where the directing staff had had its headquarters. Some two to three hundred senior officers attended, along with a sizeable contingent from Britain's ally, France. The King was the first to address the conference and limited himself to a short address in which he stressed the significance of the manoeuvres in testing the army's administrative and logistical readiness:

> It has… afforded a test of the organisation of the various branches of the Staff, and the services of maintenance, and has illustrated how greatly the operations

27 Sunderland and Webb, op. cit., p.33.
28 *The Northampton Mercury*, 26 September 1913

> of a large force depend upon administrative efficiency and good march discipline. It has also served to develop that common understanding and ready co-operation as between the different arms, units and formations, which are so essential to success in the field.[29]

It is hard to draw much from the few congratulatory comments made by the King; he was followed by Sir John French who also spoke in glowing terms. He outlined the purpose of the manoeuvres as being to assess the workings of GHQ and Army Headquarters in war conditions, whereas earlier manoeuvres were 'designed to test the capabilities of Commanders and Staffs opposed to one another'. He did emphasise the 'splendid work done by the Royal Flying Corps throughout' the manoeuvres. 'Excellent information' had been supplied. Mechanical transport 'had worked excellently throughout'.[30]

The official report, completed in early 1914 and designed for internal circulation rather than public consumption, was a good deal more detailed and critical, and prominent among its findings was the increasing impact made by aircraft. Patrick Playfair, who took part in the 1913 manoeuvres and rose to become Air Vice-Marshal Playfair, was of the opinion that 'in these manoeuvres, the position of the aeroplane was no longer that of something novel and experimental.'[31] After the White army had taken up entrenched positions south of Daventry, French employed aircraft to observe behind enemy lines, which would become the primary function of aeroplanes on the Western Front in the next few years. Notable progress from 1912 was made; it was now clear that a single aircraft could do more in reconnaissance than a large body of cavalry as long as conditions were suitable: in the single most impressive instance, one infantry division was spotted by an observer from an altitude of 7,000 feet at a distance of five miles. However, the continuing limitations of the air arm were again demonstrated: White aeroplanes failed to spot Snow's Brown 4th Division (concealed in bivouac as it had been in 1912), and on one occasion DELTA mistook some gorse bushes three miles distant for an enemy infantry brigade.

There was a recognition that these failings were due not to the aircraft so much as the inadequate training in spotting of many of the observers, who had only recently joined the RFC. There was also a significant variation between the concealment skills exhibited by units. For the first time, War Office instructions for the 1913 Manoeuvres set out ways in which troops on the ground could be concealed, though brigade commanders were enjoined to consider whether it was more advisable to continue their march and risk being spotted by aircraft or to surrender the initiative by taking cover.[32]

29 TNA WO 279/52, Report on Army Manoeuvres 1913, p.33.
30 Ibid.
31 Whitmarsh, op. cit., p.341.
32 Ibid.

Grierson's divisional commanders from the previous year, Rawlinson and Snow, successfully used concealment of troops while stationary to frustrate aerial observation, placing bivouacs and horse-lines in woods or along hedges and cooking under cover, but did not attempt to do so on the march or when engaged in action, while 1st and 2nd Divisions, part of Haig's command in 1912, attempted to do so when halted on the march and during engagements but not in bivouac – unsurprisingly 'the officers of the 1st Division... could be plainly seen at dinner.' During the Irish Command Manoeuvres, which took place at the same time, troops rushing to the side of the road at the approach of an aeroplane made themselves even more obvious, reflecting their inexperience with aircraft. The Irish Manoeuvres involved 14,000 men from 5th and 6th Divisions. The Staff of 5th Division thanked No. 2 Squadron of the RFC for 'the most excellent information which you gave us at all times, and especially as to the march of the enemy's column on the first day... within an hour of your starting, we practically knew how the whole of the enemy's forces were disposed.'[33]

Communications between aircraft and ground forces continued to be a problem, the use of wireless telegraphy on airships allowing them to report much more quickly than aeroplanes. It was clear that the reconnaissance was only of use if its findings could be communicated rapidly. The Whites had the advantage that their main landing field at Badby remained in the same location, and so they could use telephones to send out messages after the aeroplanes landed, but Brown's landing fields changed rapidly as they advanced, so they tended to rely on motorcycle and car, with much less satisfactory results. Messages were dropped to advanced headquarters at designated points marked on the ground, in order to speed up the transfer of information to front-line units, but this was not always successful. Aerial reconnaissance worked best if aircraft were equipped with wireless, but this was only the case with the airships, and even here DELTA could still only transmit messages. The airship ETA could both send and receive messages, communicating with a ground station 130 miles away and receiving signals sent by DELTA. Reflecting this new dimension in warfare, the RFC report after the manoeuvres actually raised the possibility of wireless interception and jamming.[34]

Reliability was also a genuine problem, with only a quarter of the aeroplanes serviceable after two weeks of action in exercises. No. 3 Squadron suffered an average 10% loss each day. Of its 12 aeroplanes, one was wrecked and four landed behind enemy lines. Only half of its aeroplanes were ready to fly at any given time. The squadron flew a total of 4,545 miles on reconnaissance and 3,310 miles on other flights during the manoeuvres. Its longest flight lasted 3 hours 15 minutes, during which the aeroplane travelled 190 miles. Sykes warned the War Office to expect a higher wastage rate still during a more extended campaign.

33 Ibid., p.343.
34 Ibid., p.344.

On top of mechanical problems, an added danger to aircraft was ground fire. Pilots were ordered to fly at a minimum height of 3,000 feet when exposed to rifle fire and 4,000 feet when exposed to artillery. The umpires ruled that the Brown 4th Division had shot down one of the airships on the final day of fighting, and the same division kept two howitzers specifically for anti-aircraft use, though, as *The Times* pointed out, as a howitzer could not elevate its fire to the vertical, it was ineffective in this role – anti-aircraft fire would require specialised guns. The increased danger from the ground had resulted in aircraft for the first time being painted with identification marks on the underside of their wings and an identification sheet being issued to troops before hostilities commenced.[35] This sheet consisted of woodcuts showing four types of aeroplanes in flight as seen from below as well as in front and from the side, and it was deemed to have been useful, but because new types were still being added to the RFC's strength late in the day, it wasn't possible to include all the types used. Whatever the reason, there were several instances of 'friendly fire' on aeroplanes in 1913.

Finally, there was considerable discussion as to the relative merits of aeroplanes and airships. The advantages enjoyed by the latter were acknowledged, for example the ability of DELTA to make reconnaissance flights from the early morning on 22 September in misty conditions which grounded the aeroplanes until mid-morning, but there was a general recognition that in the long run the aeroplane, being more capable of technical improvement, would be the more effective for a battlefield role.

Back in June, Seely had revealed to the King's Private Secretary 'the principal object of this year's manoeuvres, viz. the testing of the system of transport and supply.'[36] The 1913 Exercise afforded the first and only opportunity before war broke out to test fully the New Transport System introduced by the Army in 1911, which relied on the increased use of mechanised transport and reduced numbers of horses for transport and supply. Valuable lessons were learnt from the 1913 exercise, for example the potential of the motor lorry for transporting supplies and the problems of moving large numbers of transport vehicles in congested areas. By the start of the war, Britain's army was the most mechanised in Europe, and as Graham Winton has demonstrated, the BEF's use of motor vehicles in 1914 was not a response to horse shortages in the early months of the war as has sometimes been suggested but the result of planned changes which had been implemented since 1911 and which reflected the experience gained in the 1912 and 1913 Manoeuvres.[37]

Another of the elements of the support services tested in the 1913 manoeuvres, unglamorous perhaps but still vital for the smooth running of the army, was the postal system for wartime. Letters and parcels for the troops were forwarded under peace conditions by the civil postal authorities to railheads. From there they were dealt

35 Sunderland and Webb, op. cit., p.45.
36 J. Seely to Lord Stamfordham, 26 June 1913, The papers of J.E.B. Seely, 17/155.
37 G. Winton, *Theirs Not To Reason Why, Horsing the British Army 1875-1925*, pp.211-212.

with on a war basis, being despatched to refilling points in motor vehicles attached to supply columns. Field Post Offices would be established with divisional trains by the Army Postal Service, which had been set up in the same year with officers and men of the Royal Engineers Army Reserve working for the Post Office.

The APS would handle the despatch of parcels and letters and the delivery. Stamps and postal orders would be sold and postal orders cashed. Troops were instructed that letters should not contain the name of any town – 'e.g. Sergeant T. Smith, 2nd Bn., Devonshire Regiment, 5th Infantry Brigade, on Manoeuvres.'[38] Led by William Price, appointed its Director in March 1913, the APS would do the same job in the Great War, though on an unimaginably greater scale. The service tested out in 1913 was expanded in wartime to cope with hundreds of thousands of letters and parcels, keeping soldiers in touch with their loved ones at home, an important consideration in terms of the maintenance of morale. Mail was addressed to the soldiers' units and sent c/o the General Post Office in London to the BEF's field post offices, which distributed the post and also collected return mail. By October 1916 over 10 million letters and 100,000 parcels were being posted to the troops every week, the vast majority of these reaching their destination within two days of being posted.[39] 1913 had provided a vital test of the APS's capabilities.

In one respect the 1913 manoeuvres were truly innovative, as they saw a remarkable precursor of the weapon which would ultimately revolutionise land warfare, the tank. The White Army had the use of a Hornsby petrol/paraffin Crawler Tractor, which made a considerable impression as it trundled along the roads of Northamptonshire. It was photographed in Sheaf Street, Daventry, escorted by a White cyclist.[40] The Hornsby Tractor was the first vehicle which was tracked rather than fitted with wheels. It is now to be found in the Tank Museum at Bovington. Seeking to produce a vehicle for the British Army which would be able to traverse muddy or broken ground which would defeat wheeled vehicles or teams of horses, the War Office had staged a competition in 1903 in which a prize of £1,000 was offered to a tractor capable of hauling a 25 tonne load for 40 miles without stopping for fuel or water; the 80 horsepower, 12 tonne Hornsby tractor had been the only entrant to complete the course and had managed to go a further 18 miles before running out of fuel. When soldiers first witnessed this vehicle, now equipped with tracks, at further trials in July 1907, they coined the word 'caterpillar' to describe the machine. King Edward VII met Hornsby's Managing Director David Roberts in May 1908 and saw the machine in action at Aldershot. The British Army, keen to move away from the use of steam traction engines for towing artillery, ordered four of the new caterpillar tractors in 1910, and it was one of these which was present at the 1913 manoeuvres. As early as 1910 the feasibility of mounting a gun and bullet-proof shields on the vehicle was being

38 NA WO 279/52 Report on Army Manoeuvres 1913 p.12.
39 Andrew Rawson, *The British Army 1914-1918*, p.157-158.
40 Sunderland and Webb, op. cit., p. 64.

discussed. Unfortunately, officers from the Royal Artillery were unimpressed with the tractors, complaining of their noise and slow speed; as one Battery Commander wrote, 'The team of eight horses in my opinion is far superior under every condition.'[41] The Director of Artillery, Brigadier-General Stanley von Donop, queried the whole purpose of the vehicles and put a stop to all trials.

These machines had originally been developed in Lincolnshire for use in farming, but when the expected orders from civilian customers also failed to materialise, Hornsbys sold the patent in 1914 to the American firm Holt Manufacturing (now Caterpillar Inc., the world's leading maker of construction equipment). Within a year of the start of the war the conditions of warfare in France had persuaded the British Army to order 442 of Holt's caterpillar tracked vehicles, made under licence by Ruston in Lincoln. In addition to the massive contribution the Holt Tractor made to the Allied war effort, mainly hauling medium guns such as the 6-inch howitzer, the 60-pounder, and later the 9.2-inch howitzer, it also contributed to the British development of the tank. It was the primary influence on a key figure in the tank's development, Lieutenant-Colonel Ernest Swinton, even if the contract was ultimately given to another Lincolnshire company, Fosters & Sons and its managing director and designer, Sir William Tritton. Holt components provided the basis for early French tanks, the Schneider and Saint-Chamond, and tractors commandeered by the Austro-Hungarian army and loaned to the Germans formed the basis of the German A7V tank.

The use in the 1913 exercise of such a new and ultimately significant invention belies the reputation of the pre-war British Army for being reluctant to embrace new technology, in much the same way as does its enthusiastic if belated discovery of aeroplanes. It is easy to criticise the British Army for failing fully to see the potential of tracked vehicles before the war, but it is worth remembering that German generals also examined and considered the Holt Tractor just before the Great War before rejecting it. The British were also concerned about the risks in having such a volatile fuel in close proximity to artillery shells.

In many respects the 1913 exercise reflected an army which had now come to grips with manoeuvres. There were fewer complaints about umpiring decisions and fewer instances of unrealistic outcomes of combat. To a large extent this reflected a modified umpiring system. In a sincere attempt to improve on 1912, revised instructions for umpires had been issued in June 1913 which sought to emphasise the responsibility of regimental officers for the practical conduct of operations and the avoidance of unrealistic situations. This was accompanied by a new reporting system, which largely removed the burden from the umpires and passed the task on to the narrative officers. The only written reports for umpires in units smaller than a division would now be reports of any decisions they gave. One of the most frequently heard complaints after

41 For the Hornsby tractor and its part in the development of the tank, see J. Glanfield, *The Devil's Chariots: The origins and secret battles of tanks in the First World War*.

Hornsby Tractor, used by the army in the 1913 Army Exercise. (Alamy)

manoeuvres was that inflated demands for compensation were submitted by local landowners. In 1913 there were some complaints that crops had been damaged, haystacks had been nibbled by horses and farm equipment had gone missing, but generally the Chief Compensation Officer, Major-General Sir John Hanbury-Williams, based in the Pomfret Hotel in Towcester, complimented the district's landowners and farmers on the moderation of their compensation claims.[42] When taken with the success of the Army's transport and supply arrangements, identified beforehand as its main purpose, there were grounds for judging the 1913 exercise as a success.

However, in one important respect the 1913 Manoeuvres revealed serious systemic problems in the British Army, and these were to be found at the very top. These issues would have serious implications for the command structure of the British Expeditionary Force which was sent to the continent the following year. French had been assured in November 1912 that he would be the Commander-in-Chief in the event of war, but many of those at the top of the British Army had little faith in his suitability.[43] These reservations were only reinforced by the events of September 1913. Haig also came in for some criticism for allowing a three-mile gap to develop between his two divisions in the advance before the final battle, a recurrent problem

42 NA WO 279/52 Report on Army Manoeuvres 1913 p.6.
43 Walter Reid, *Douglas Haig, Architect of Victory*, p.272.

for generals unused to commanding such large bodies of troops and one which would come back to haunt the BEF in August 1914, when a gap of fifteen miles opened up between Haig's and Smith-Dorrien's corps after Le Cateau. The main criticism, though, was directed at the man who would be leading the BEF on the continent within a year, Sir John French.

This was not the first time that French had stumbled when faced with the apparently straightforward task of dealing with a skeleton army. During the 1907 Aldershot manoeuvres, he had commanded an attacking force in a two-day exercise which involved a defending force commanded by Lieutenant-Colonel Hubert Gough. This was a flagged exercise, where units were represented by a few soldiers with flags to signify the edges of the body of troops. In a letter to his father, Gough remarked that 'As a matter of fact I don't think it was a very difficult task to defeat Sir John.' Gough's criticisms of French's performance included his failure to use his artillery effectively in combination with his infantry, the fact that he dispersed his forces over too wide a front and his decision not to employ a reserve.[44] As a result, Gough was able to deal with French's piecemeal attacks and then launch an offensive of his own. Such a performance was hardly likely to instil much confidence among commanders in the man who would lead them to war when it came.

Repington in *The Times* was extremely critical of French's performance in September 1913 and ridiculed the dual role occupied by him, as he had been not only the Brown commander but also the Director of Manoeuvres.[45] French was ready to admit that 'the manoeuvres taught *us all* many lessons' but was hurt by Repington's suggestion that he had struggled to beat a skeleton army, dismissing his old friend's remarks as 'childish, *stupid*, and inclined to be rancorous'.[46] Repington was insistent in his refusal to back down from his criticism of French's performance, based as it was on information he had received from three of the Field Marshal's senior commanders, Haig, Rawlinson and Lawson as well as William Robertson and Forestier-Walker, Chief of Staff to Paget's 2nd Army. Repington pointed out to his editor that many generals who had attended the manoeuvres would have been far more critical than he, and he insisted that he would stick to his views 'even at the cost of the loss of a valued and old friendship'.[47]

It was not just French's relationship with Repington which was damaged by the events of September 1913; he also had a falling-out with his Chief of Staff, James Grierson, the star of the previous year's manoeuvres. Ever since January 1906 Grierson had expected to be Chief of Staff in the event of war, recording in his diary a 'long talk with French in my office… He promised to take me as Chief of the General Staff if he

44 Stephen Badsey, *Doctrine and Reform in the British Cavalry 1880-1918*, p.200.
45 *The Times*, 10 October 1913.
46 R. Holmes, *The Little Field-Marshal: Sir John French*, p.149.
47 The Letters of Lieutenant-Colonel Charles a Court Repington CMG Selected and Edited by AJA Morris, p.212.

commands.'[48] This had been confirmed by a further discussion against the backdrop of the Morocco Crisis in the summer of 1911, when 'French talked to me of war in which I should be his CGS and Douglas and Paget command armies under him'.[49] However, French was unimpressed by Grierson's performance in this role during the 1913 manoeuvres, and even before he left Northamptonshire he had told Henry Wilson that he had changed his mind about the post of Chief of Staff. Grierson was to be replaced by Sir Archibald Murray ('a complete nonentity' according to Edmonds).[50]

Grierson was instead given command of 2nd Army in place of Paget (who had also fallen out with the fractious French in 1913), an appointment 'which is a great thing, and which I much prefer to CGS'.[51] Haig, designated to command 1st Army, sided with Grierson, recording that French's orders 'were of such an unpractical nature that his Chief of the General Staff (Grierson) demurred'.[52] The suspicion must be that French preferred a malleable Chief of Staff such as Murray to one who would stand up to him, though it is also the case that Murray had a strong claim as an experienced staff officer who had occupied the same role for French before, for example at the 1906 manoeuvres.[53] There is, in any case, no contemporary evidence to suggest that Murray was French's first choice as Chief of Staff in August 1914, as he may have preferred Wilson.[54]

Whatever the case, there was a widespread feeling that Grierson had been harshly treated. Repington had praised his work as Chief of Staff– 'General Grierson deserves the greatest credit for his organisation which, by its completeness and attention to the minutest details, enabled the changes of front to be made with perfect smoothness.'[55] This encomium could, of course, have been prompted by a desire on Repington's part to defend his old friend from French's complaints, but French's response to criticism – to put the blame on his subordinates – did not bode well for the future. According to John Charteris, Haig 'had noticed in manoeuvres that Sir John French did not appear to have grasped the conditions of modern warfare, and that he was impatient of advice'.[56]

Once the soldiers had moved on, and with them the press and visitors, life returned to normal in rural Buckinghamshire and Northamptonshire. To some the adjustment was a hard one to make, so easy had it been to be swept up in the excitement of recent events, as one school newsletter suggests: 'We are all military this week. We can't help it as the feeling has been forced on us.'[57] For many local people, the most signifi-

48 D. Macdiarmid, *The Life of Lieut. General Sir James Moncrieff Grierson,* p.216.
49 Macdiarmid, op. cit., p.241.
50 N. Gardner, *Trial by Fire: Command and the British Expeditionary Force in 1914*, p.5.
51 Macdiarmid, op. cit., p.256.
52 Lt-Gen D. Haig (ed. G. Sheffield and J. Bourne), *War Diaries and Letters 1914-1918*, p.58.
53 *The Evening News*, 25 September 1906.
54 S. Badsey, 'Sir John French and Command of the BEF' in S. Jones, *Stemming the Tide* p.41.
55 *The Times*, 27 September 1913.
56 Brig-Gen. J. Charteris, *Field-Marshal Earl Haig*, p82.
57 Bugbrooke and the Great War, http://www.bugbrookelink.co.uk/WW1 (accessed 30/8/2015).

cant feature of the week had been the presence of the King, an unsurprising fact in a patriotic country where opportunities to see their monarch in the flesh would have been rare, at least in a Northamptonshire village. One eyewitness in Eydon left the following account:

> He rode a beautiful black horse and looked just like the picture we'd seen of him. When he came by Henry lifted his bowler hat and held it above his head, and we took our caps off and stood very quiet-like, when old Anne shouts out 'God bless King George!' The King smiled and held up his hand… as she stood on her doorstep with a little black bonnet on her head and a big coarse apron tied around where her waist ought to have been.

The observations of local clergymen are especially informative. The Rev. F.S. Keysell of Weedon, writing in his Parish Magazine for October 1913, described the manoeuvres as 'the one absorbing object of interest', especially among the young men and those older men who had themselves once served in the Army. The Rev. Frank Churchill of Everdon's parish had been 'a scene of much life and interest during this month.' On 25 September, 'aeroplanes descended and ascended at intervals in Mr Goodman's field between Great and Little Everdon, and Everdon was bombarded by Artillery and Infantry and the din of battle raged during the whole morning.' The vicar observed, 'From what we have witnessed we have some idea of what our soldiers have to go through for the defence and honour of their country for which they surely deserve our respect and sympathy.'[58]

The village of Bugbrooke, seven miles south-west of Northampton, was another village right in the middle of events. The newsletter of the village school recorded on Friday 19 September: 'As far as we can understand we are now between two big armies, the one northward, and the other southward, and today (Friday) one part of the North Army will be in Bugbrooke at about four o'clock this afternoon. About 900 men will bivouac in the park, and tomorrow sometime we shall probably see something of the fight. Fancy quiet little Bugbrooke being part of a great battlefield!' More than 2,000 troops passed through the village the next day, and the airship DELTA passed overhead on the 22nd.[59] When it was all over, and calm had returned, the Rev. Ernest Harrison, Parson of Bugbrooke, reflected in his parish magazine on what the village had witnessed:

> It has been one of the most exciting weeks in Bugbrooke history and interesting in the extreme. But has it no lessons for us to learn? Surely to put more force and meaning into our prayer, "Give peace in our time, O Lord." If war came,

58 Sunderland and Webb, op. cit., p.48.

59 Bugbrooke and the Great War, http://www.bugbrookelink.co.uk/WW1 (accessed 30/8/2015).

> Farthingstone, Litchborough, Stowe, Everdon, Dodford, all parishes in this Deanery would now be heaps of smouldering ruins, the lovely country, hills and valleys and woodland full of dead and dying, the civilian population, women and children in particular flying as best they can … along roads more and more blocked with panic stricken fugitives, homeless, foodless, without shelter of the least description.[60]

Within a year the Belgian cities of Dinant and Louvain lay in ruins, the Battles of the Frontiers had claimed hundreds of thousands of casualties and the roads of Belgium and northern France were clogged with a stream of refugees displaced in the opening month of the war. Safe from the fighting on the continent, Bugbrooke would be spared the devastation inflicted during the war on so many villages in Flanders, but the war memorial in its beautiful parish church bears the names of 27 men of the village who perished in the war whose effects the Rev. Harrison so accurately foretold. Two of the names are of soldiers who took part in the 1913 manoeuvres – Lance-Corporal (Walter) James Clarke of 1st Battalion, Northamptonshire Regiment, was part of the guard provided at Althorp for the King's visit while his older brother, Private Mark Clarke, served in the same battalion, part of 8th Infantry Brigade, in Rawlinson's division. Within a little over a year, James had been killed at Ypres. Mark would be killed in July 1916 on the Somme, still serving in the Northamptonshire Regiment as part of Rawlinson's Fourth Army.[61]

60 Sunderland and Webb, op. cit., p.48.
61 Bugbrooke and the Great War, http://www.bugbrookelink.co.uk/WW1 (accessed 30/8/2015).

## 6

# Foreign Manoeuvres, Foreign Wars

As well as staging their own exercises and manoeuvres, the British Army could also learn valuable lessons by sending observers to the manoeuvres held by the armies of other countries, whether these were allies, neutral states or potential enemies. In addition, British officers often used up their leave in visits to the battlefields of the Napoleonic Wars and Franco-Prussian War. Jimmy Grierson visited the Napoleonic fields of Leipzig (1884), Waterloo (1889) and Bautzen (1896) as well as accompanying Sam Lomax on a tour of the Peninsular War battlefields in 1911. When the 1911 manoeuvres were cancelled, he spent the time touring the French frontier with Sir John French and the French Military Attaché, Colonel Huguet. In the same year, Henry Wilson spent a week in Belgium cycling around the area where he expected the BEF to deploy when war came, and then headed south to Verdun, Nancy and the 1870 battlefield of Mars-la-Tour. Wilson had first visited the battlefields of the Franco-Prussian War in April 1893 when at the Staff College; others on the same tour, led by their lecturer G.F. Henderson, included Rawlinson, Snow, Haldane and Byng. The following year Wilson and Rawlinson returned to the same battlefields, this time on a cycling tour.[1] The regular flow of officers to the continent on battlefield tours is another sign of the increasing seriousness about their profession to be found in the British officer corps as war approached. Further afield, Thompson Capper led a party of 20 instructors and students from the Indian Staff College on a three month tour of the battlefields of Manchuria in 1907, gaining direct experience from the most recent and relevant of wars.[2]

Many officers chose to visit foreign manoeuvres in a private capacity. While at the Staff College Tom Bridges spent all his leave abroad, going to foreign manoeuvres and on battlefield tours, including the Shendandoah Valley, inspired by Henderson's

1 K. Jeffrey, *Field Marshal Sir Henry Wilson: A Political Soldier* p.19.
2 J.P. Harris, *The Men Who Planned the War: A Study of the Staff of the British Army on the Western Front, 1914-1918*, p.37.

lectures on Stonewall Jackson.[3] At the French manoeuvres in September 1906 the official British party encountered a group of officers, including Rawlinson and Du Cane, attending in civilian clothes.[4] Nor was the practice of visiting foreign manoeuvres restricted to regular officers; Colonel Henry May of the Artists Rifles, the 28th (County of London) Battalion of the London Regiment, attended continental army manoeuvres in civilian dress, with War Office permission. A solicitor by profession, May was joined by two other officers from his territorial battalion when he attended the Dutch (1911), French (1912) and Swiss (1913) manoeuvres, and they were scheduled to visit the German in 1914 before the outbreak of war intervened. May later recalled that:

> We saw a lot of the foreign military work, and I learned a lot of little things and different methods and of how they managed military affairs on the Continent that were very useful to me afterwards when war actually broke out, and we found ourselves soldiering on the Continent in grim earnest.[5]

In one year alone (1912) the British sent observers to the manoeuvres held by the armies of their ally Japan, the two Entente powers (France and Russia), all three members of the Triple Alliance (Germany, Austria-Hungary and Italy) and six neutral powers (Denmark, Greece, Norway, Sweden, Switzerland and the United States).[6] There were, of course, some limits to what foreign observers were allowed to see, but fewer than might be expected. Hosts were surprisingly open and hospitable with representatives of potential enemy nations, at least in the early years of the century. Ominously, by 1912 the British army was lamenting the fact that 'the opportunities for observation afforded to officers attending foreign manoeuvres are becoming more and more limited.'[7] This reflected an increasingly tense international climate, with the development of two ever more clearly defined armed camps and a series of war scares of which the Morocco Crisis of 1911 was only the latest. Spy scares, the outstanding example of which was the coming to light in April 1913 of the Redl scandal in Austria-Hungary, made hosts reluctant to reveal too much, for example their airships or latest field guns, to foreign visitors. Colonel Frederick Trench, Military Attaché in Berlin between 1906 and 1910, reported to his superiors that 'the movements of strangers are more closely watched by the police both at and away from manoeuvres than has previously been the case.'[8] The British also noted that more and more of the time of military attachés attending German manoeuvres was taken up with an elaborate

3 Lt. Gen. Sir Tom Bridges, *Alarms and Excursions – Reminiscences of a Soldier*, p.55.
4 TNA WO 27/507, Report on French Manoeuvres 1906, 3 September 1906.
5 Col. H.A.R. May, *Memories of the Artists Rifles*, p.69.
6 TNA WO 33/618 Report on Foreign Manoeuvres 1912.
7 Ibid.
8 M. Seligmann, *Spies in Uniform: British Military and Naval Intelligence on the Eve of the First World War*, p. 172.

round of social events which took them away from the field of operations. During the 1912 Imperial manoeuvres, an entire morning was taken up with a tour of a medieval castle. The staff officers selected to escort foreign attachés were increasingly used to prevent them from getting too close to the action. Trench complained that the officer accompanying the attachés at the 1907 manoeuvres 'apparently interpreted his instructions to mean that he was not to let anyone of them out of his sight or to afford any information except of a trivial kind.'[9]

In any case, foreign armies did have some cause to be suspicious of their British guests. While Imperial directives of 1878, 1890 and 1900 cautioned German military attachés against illegal acts of intelligence gathering, the British had fewer qualms. Brackenbury as Head of Britain's Intelligence Department from 1895 was keen to use every means at his disposal, though he was reluctant to rely on attachés as they were often closely watched and would find it hard to deny charges of spying thanks to their official status. Brackenbury preferred to use officers who were visiting while on leave. One such officer was Robert Baden-Powell, who once posed as a painter, hiding in dots and dashes on his canvas an outline of Algerian coastal defences. He also managed to gain access to German manoeuvres, thereby getting close to a machine-gun being tested. Amateur spies could, though, cause embarrassment, for example the officer who was accosted by two gendarmes and proceeded to start eating his notebook, whereupon he was arrested. According to Count Gleichen, they had been asking him for a light. Brackenbury made it clear that such individuals 'were not to consider themselves as employed by the Intelligence Division, and are, on no account whatever, to present themselves to any person as being so employed, or as being engaged in any official work, or as having any official mission.'[10]

The British noted marked differences in the approaches adopted by the various countries in their approach to manoeuvres. Japan and, to a lesser extent, Germany, were observed to carry out manoeuvres according to a carefully pre-arranged programme; 'in France, on the other hand, considerable latitude is given to commanders.'[11] Similarly, the French allowed their umpires some freedom of action whereas the Russians provided elaborate and prescriptive instructions for theirs. One valuable observation which the British failed to act upon was that the French allotted their umpires to different zones, which changed daily, rather than to specific units, thus reducing the danger of partiality in their decision-making.

As well as a natural tendency towards chauvinism – every army felt that their methods worked best for them – one of the inhibiting factors for British observers in terms of learning from other countries was a feeling that many of them were not using manoeuvres in the same way and for the same purposes. In some of the armies

9 Seligmann, op. cit., p.105.

10 C. Brice, *The Thinking Man's Soldier: The Life and Career of Sir Henry Brackenbury 1837–1914*, pp.179-180.

11 TNA, WO 33/618, Report on Foreign Manoeuvres 1912, p.iv.

observed, the transport, medical and other services did not participate in manoeuvres, preferring to carry out field exercises earlier in the year. There were also regular complaints about a lack of realism: 'In several cases... army manoeuvres are designed for spectacular effect, and even in France where they appear to be most realistic, many details of procedure such as ranging by artillery and use of cover by infantry are not carried out.'[12]

German army manoeuvres were the main focus of attention for British officers, especially after 1904 when it became increasingly obvious that a future war would pit the two nations against each other. Grierson, as Military Attaché in Berlin between 1896 and 1900 and later as DMO at the War Office (1904-1906), was a regular attender at the manoeuvres held by Germany (1889, 1890, 1892, 1896-1899 and 1905) and France (1892, 1906, 1909) as well as those of Austria-Hungary (1890) and Belgium (1906). He was struck by the warm reception he was accorded by his German hosts, and in particular by the Kaiser himself, which can be explained both by Grierson's personal charm and by the All-Highest's childish delight at his discovery that the two men had been born on the same day. By 1898, however, Grierson's view of the Germans had undergone a dramatic reversal, even if he continued to get on well with them socially. He detected a changed attitude in Berlin towards Britain and by 1898 admitted to a friend in the Intelligence Division that 'I am sick of it and longing to be "back to the Army again."'[13] To the same correspondent he revealed his personal view that: 'We must go for the Germans and that right soon or they will go for us later. A pretext for war would not be difficult to find', a remarkable statement given Britain's diplomatic isolation at the time. Grierson's detailed reports to the Intelligence Department after each set of manoeuvres played a vital part in forming the increasingly hostile British view of the German Army.

Later attachés in Berlin found it hard to emulate Grierson's delicate balancing act between outward *politesse* and inner revulsion. As the cousin of the Kaiserin Augusta Viktoria and a member of a German princely house, Count Gleichen (Military Attaché 1903-1906) had been expected to get on well with his hosts, but he was given a frosty reception, widely being seen as a traitor to his German roots, and he angered the Kaiser by making unauthorised visits to the capitals of those German states which had their own armies.

British observers at German manoeuvres tended to be impressed by the size and scale of what they had seen rather than the tactical prowess on display. Grierson observed in 1899 that 'admirably drilled, disciplined and trained to march and shoot as the German infantry is, I imagine that I see a distinctly retrograde tendency in its tactical handling.' 1899 marked the first appearance of machine guns attached to battalions on a German manoeuvre ground and Grierson observed that 'as in all

12 Ibid.

13 D. Macdiarmid, *The Life of Lieut. General Sir James Moncrieff Grierson*, p.116.

novelties the Germans have yet to learn the proper tactical use of the Maxim.' [14] It was a similar story a decade later: Aylmer Haldane reflected that 'the impression I took away from what I saw was that if we had to fight the German Army we might expect it to continue to follow its antiquated attack formations, and very costly they would prove to be.' Haldane felt that 'in spite of what their attachés had seen in South Africa and Manchuria the lessons had not been taken to heart.'[15]

This was also the case with Winston Churchill, who attended in 1906 and 1909. After the 1906 manoeuvres in Silesia, he wrote to his chief Lord Elgin, Secretary of State for the Colonies, to report that 'I do not think they have appreciated the terrible power of the weapons they hold & modern fire conditions, and have in that & in minor respects much to learn from our army.'[16] The comments of Charles Repington in *The Times* on the 1911 German manoeuvres are significant for repeating many of the same charges which were frequently levelled against the British Army:

> The infantry lack dash, display no knowledge of the ground, are extremely slow in their movements, offer vulnerable targets at medium ranges, ignore the service of security, perform the approach marches in old-time manner, are not trained to understand the connection between fire and movement and seem totally unaware of the effect of modern fire.[17]

However, there was also an objective recognition among British visitors that the Germans were becoming increasingly adept at fighting the new type of war. Infantry assaults were developing in a more measured way. The dispersed battle order was becoming the new orthodoxy, with divisions and corps spread out along broader fronts. Machine gun detachments were working more closely with infantry, and cavalry were becoming accustomed to scouting and auxiliary attack roles rather than being deployed in a shock role. Rapid-firing guns and howitzers now operated from hidden positions rather than up in the front line as before. The 1909 and 1910 manoeuvres even featured airships, and then aeroplanes in 1911. Under von Moltke, 'there was now a much more concerted effort to use manoeuvres to wean corps commanders from practices and techniques that did not conform to the demands of modernity.'[18] Based on their experience of German exercises and manoeuvres in the decade before 1914, British generals and politicians were well aware of the threat posed by the German army. After a second visit to the Imperial Manoeuvres in 1909, Churchill reported 'an enormous advance from 1906' and wrote to his wife Clementine that 'this army is a terrible engine.'[19]

14 Ibid., p.141.
15 General Sir Aylmer Haldane, *A Soldier's Saga*, p.258.
16 D. Russell, *Winston Churchill: Soldier: The Military Life of a Gentleman at War*, p.334.
17 M. Adkin, *The Western Front Companion*, p.135
18 E. Dorn Brose, *The Kaiser's Army: The Politics of Military Technology in Germany during the Machine Age, 1870-1918*, p.157.
19 D. Russell, op. cit., p.336.

Apart from Germany, Britain's likely enemy in the event of war, the most significant sets of manoeuvres were those held by France, especially after the signing of the Entente between the two nations in 1904. The length at which French Army manoeuvres were discussed in official reports (41 pages in the 1912 report) and the extensive coverage in *The Times* reflect the developing relationship between the two countries, cemented by regular visits by Sir John French, Grierson and Wilson in particular. As well as attending the French manoeuvres in 1906, 1908 and 1911, Sir John hosted a group of French officers, led by General Michel and General Marion, at Aldershot in 1907. Having witnessed the Aldershot Command's divisional manoeuvres, the two generals returned to France with, in the words of Colonel Huguet, 'a knowledge of your Army far greater than they had before. They can now appreciate its high training and see what British troops, led by such officers as those of the Aldershot Command, are able to do.'[20] Wilson was a regular attender at French manoeuvres after his appointment as DMO in 1910, and the pace picked up after the Morocco crisis of 1911. At the same time as Foch was attending the 1912 manoeuvres in East Anglia, Wilson was with the French Army in Touraine. Castelnau attended the 1913 manoeuvres, where he extended an invitation to Ivor Maxse to spend some time with the French Army, which resulted in Maxse and Wilson attending a staff tour in May 1914 which started in Paris and then moved to the border with German-held Lorraine.[21] Joffre was due to attend the 1914 manoeuvres which never happened because of the outbreak of war.

Other British officers frequently attended manoeuvres as well as field exercises, musketry meetings at Mailly and Châlons or special conferences; the last such conference, which took place at Mailly in July 1914 was attended by Grierson, Haig and Allenby and was designed to familiarise the British General Staff with the workings of the French railways. Theory was studied in Paris, before moving on to Amiens for a practical demonstration of concentration, supply and the principles of railway movement in war.[22] In the first week of July, with the crisis following the assassination of the Archduke Franz Ferdinand developing in Vienna and Berlin, Grierson and his colleagues observed exercises carried out by the French 11th Division and consulted with Joffre, Foch and Castelnau. Within a month they would be back in France at the head of an expeditionary force sent to assist their allies. Grierson recorded in his diary that 'altogether the 11th Division is in a splendid state of efficiency'.[23] The division's commander, General Maurice Balfourier, would go on to play a leading part alongside the British in the Somme campaign in command of XX Corps.

Such was the warmth of the relationship between those at the top of the two armies, typified by the friendship between Wilson and Foch, that the French high command enjoyed a much stronger bond with the British than they did with their Russian allies,

20 D. Macdiarmid, op. cit., pp.234-235.
21 J. Baynes, *Far From a Donkey: The Life of General Sir Ivor Maxse*, pp.106-107.
22 General V. Huguet, *Britain and the War, A French Indictment*, pp.26-27.
23 D. Macdiarmid, op. cit., p.254.

who were reluctant to disclose much when Joffre and Foch attended their manoeuvres in 1910.[24] This was despite the fact that the French were certain of Russian assistance in the event of war whereas they were much less confident that they could count on a British force joining them on the continent. Ironically, the Franco-Russian alliance was set in stone but lacked detailed planning while the plans for moving the BEF across the Channel were all in place by August 1914, thanks in large measure to the impetus provided by Henry Wilson. As Samuel R. Williamson has pointed out, 'Anglo-French co-ordination far exceeded that established between Paris and St Petersburg.'[25]

British views on French manoeuvres tended to be equivocal. Officers generally found the infantry's marching and powers of endurance highly impressive and were delighted to have the opportunity to see the matchless 75mm gun in action. British artillerymen at the 1910 French artillery manoeuvres in Picardy were impressed with what they saw and this led to an important debate at the next General Staff annual conference. While they were keen to learn from the French, especially in terms of the way their rapid-firing guns served the needs of the infantry, there was a recognition that the French example could not simply be adopted wholesale as they had a different conception of the role of artillery in a modern battle and grouped their guns in four, not six, gun batteries as the British did.[26]

If the French field guns made a great impression on British observers, they were less impressed with the cavalry's care for its horses and were concerned at the almost total lack of field howitzers and heavy field artillery. They also noted that the French employed dense infantry formations to attack, and made little attempt to use cover. British officers attending the 1912 manoeuvres confessed themselves at a loss to work out the system underlying them.[27]

These observations provide an interesting contrast with the views of the Germans, as seen in their intelligence assessments. Of the 1908 manoeuvres, when the rival forces were commanded by Generals Millet and Trémeau (the designated Commander in Chief in the event of war) the German report observed: 'As in previous years, the French commanders have displayed exaggerated caution and limited initiative. Both commanders covered themselves against all possible dangers and subordinated their decisions to the anticipated actions of the enemy.' In 1911, when General Besancon was opposed to General Chomer, the Germans remarked that 'the leadership of both the commanders in this manoeuvre was more careful than daring.'[28] This persistent German criticism of the caution displayed by French generals was not something

24 E. Greenhalgh, *The French army and the First World War*, p.23.
25 Ibid., p.25.
26 Edward M Spiers, 'Rearming the Edwardian Artillery', *Journal of the Society for Army Historical Research*, pp. 172-176.
27 D. Porch, *The March to the Marne: The French Army 1871-1914*, p.214.
28 R. Foley, 'Easy Target or Invincible Enemy? German Intelligence Assessments of France before the Great War', *Journal of Intelligence History*, p.8.

which occurred to British observers. It may be explicable because of the culture of heroic cavalry actions inculcated by years of Kaiser Wilhelm's interference in German manoeuvres, but it also calls into question easy assumptions about the French army's supposed commitment to the *attaque à outrance* and the cult of the offensive.

As well as French tactics and leadership, German army reports were also critical of the French troops, above all their fire discipline. French infantrymen were seen as too temperamental. As von Moltke wrote in October 1912, 'The Frenchman is an intelligent and competent soldier, who fervently loves his motherland and who is capable of being moved to great exertion. However, he is nervous and his morale quickly collapses... In a defeat, it is difficult to maintain order.'[29] There was widespread acknowledgement, however, that in technical areas, above all field artillery and in the air, the French led the way.

For the British, the two most important sets of French manoeuvres before the outbreak of war were those of 1906 and 1912. The 1906 manoeuvres involved 33,000 men of the 2nd Corps under the direction of its commander, General Victor-Constant Michel. They took place between the Marne and the Aisne within the quadrilateral formed by the towns of Compiègne, Soissons, Senlis and Chateau Thierry, and so afforded an opportunity to practise on the future battlefields of September 1914. Sir John French was accompanied by Grierson, Lieutenant-Colonel Lowther (the British Military Attaché to Paris) and Colonel Huguet, his French counterpart. Also in attendance was French's ADC Lieutenant Maurice Brett, whose significance lay less in the detailed and highly entertaining diary he kept between 30 August and 9 September 1906 than in his being the second son of Lord Esher, the highly influential courtier who had played such a large part in the reform of the British army after the Boer War and who was still keen to keep an eye on his protégé French – in the same way he found his eldest son Oliver a post as secretary to John Morley, the Secretary of State for India.[30]

Having crossed to France early on 31 August and dined at the Ritz that evening, the British officers left Paris for the manoeuvres from the Gare du Nord on Saturday 1 September. Sir John French's train stopped en route at Villers-Cottérêts and Crèpy en Valois, both of which would witness British rearguard actions in the closing days of his army's retreat from Mons eight years later. That evening he and the rest of his party attended a dinner for the foreign observers at the Chateau of Compiègne hosted by General Michel, at which the warmth of the reception afforded to the British military mission was seen to reflect increasing co-operation between France and Britain.

29 Ibid., p.9.
30 Brett's diary is contained in the War Office report on the 1906 French manoeuvres, which also contains the handwritten script of French's speech of thanks at the reception on 1 September among many other items of interest. TNA WO 27/507, Report on French Manoeuvres 1906.

Speaking to a large audience which included German, Russian and Austrian observers as well as British, Michel addressed Sir John French directly: 'Your presence among us is a new proof of the sympathy which your gracious sovereign has always shown towards France and of the cordial relations which unite our two countries.' Sir John responded with a gracious and well-received speech delivered in French which suggests that his facility in that language was perhaps greater than has sometimes been appreciated, at least when reading from a script. While clearly not as proficient as Wilson or Grierson, it seems that he could read French and follow a conversation. Very few French generals could converse in English – certainly Joffre and Foch could not – and British officers seemed to pride themselves on their limited linguistic grasp. According to Edmonds, Charles Townshend, who had married a Belgian lady, was given to dropping French expressions into his comments at conferences. He once spoke of Le Coup de Marteau (a hammer blow) being delivered, at which a loud voice was heard to call out 'What's he saying about tomatoes?'[31]

The cosmopolitan Brett, who would spend most of the coming war as a liaison officer in Paris, was able to see for himself the practical problems entailed by the Entente Cordiale when he and Sir John returned to the capital at the end of the manoeuvres. At a dinner at the Club du Rue Royale, Sir John and General François de Négrier enjoyed a 'most interesting talk about army and politics' but, Brett recorded, it proved 'difficult to make each other understand'. The French general's views on the subject are likely to have been robust; it was only two months since the Dreyfus Affair had ended, officially at least, with Dreyfus' rehabilitation and reinstatement, and barely a month since de Négrier had fought a duel with former Minister of War Louis André.

There were 30 observers from 23 countries present at the 1906 French manoeuvres, including Germany, Russia, Austria-Hungary, Serbia, Belgium, Japan, Spain, Chile and Mexico, but only the British stayed in the field with the 2nd Army Corps while all the rest stayed in Compiègne, and only the British were invited to attend the briefing at the end of each day's proceedings. Brett's diary suggests that the French troops left a good impression on their visitors. He noted that 'from the physical point of view they are not much shorter than our men. Cheerful, willing and extraordinarily good marchers.'

On the final day of the manoeuvres, Saturday 8 September, Brett reported a 'splendid cavalry charge… infantry came on in great thick heavy lines, well supported by guns. Very interesting.' Overall, though, he noted that 'they do not seem to know how to use their cavalry', preferring to make a frontal assault against the defensive position rather than working around the enemy's flanks or attacking their line of communications. Brett thought the artillery to be extremely well-handled but was surprised by the very dense infantry formations and the fact that no use was made of the ground for cover. He also observed that 'their trenches are

31 Brig.-Gen. J. Edmonds, *Memoirs* chapter XXII, p.21.

Men of the 67th Regiment at the 1906 French manoeuvres. (Author's collection)

useless.' The horses were 'small but healthy'. One matter of concern was the heavy burden of the French infantryman's knapsack; Brett noted that 'they take off their knapsacks whilst firing and put them on again when they move. If they were in hot weather they would be certain to leave them behind.' General Michel's professionalism and calm common sense made a profound impression on the British party, and his status as the coming man in the French army was confirmed by his appointment in January 1911 to the post of Vice-President of the Conseil Supérieur de la Guerre, which marked him out as the designated Commander in Chief of the Army in the event of war. However, Michel's proposals for a defensive strategy, based on the assumption that the main German thrust would be aimed at Belgium, were politically unacceptable, and led to his dismissal by War Minister Messimy in July 1911, at the height of the Agadir Crisis, and his replacement by the more offensive-minded General Joffre.

After a meeting in Paris with President Poincaré, who did not leave a favourable impression – 'a man with a beard, no presence and no personality' – Brett returned to Britain along with his chief. His abiding impression of the French was that 'every soldier seems proud of his army and confident that next time they will beat the Germans. They only wish it would come at once instead of the delay strengthening the German position.'

Grierson had arrived in Paris having previously attended the Belgian manoeuvres in the Ardennes, and was struck by the comparison between the two armies. He thought that the Belgian troops were 'unintelligent looking, slack, and I should say not well instructed.' March discipline was 'decidedly defective as regards the formation and the movements of the column' while the artillery was 'quite behind the age'.[32] In contrast, having not attended French manoeuvres for over ten years he was greatly struck by the enormous improvement in every respect in their army. Grierson was especially impressed by the staff arrangements, the administrative services, ammunition columns, and hospitals. All in all, he felt that the French Army's 'spirit is excellent, its leadership skilful, and its preparation for war thorough and complete'. Grierson believed that if it could win the first battles of the next war and regain the confidence lost in 1870, 'the French Army will be hard to stop in its career of success.'[33] Grierson's view may have been somewhat over optimistic, but the 1906 manoeuvres undoubtedly played an important part in fostering good relations at a time when the military conversations between the two countries were at a crucial stage.

The 1912 manoeuvres, which took place in the Loire Valley between 11 and 13 September, were attended by the Grand Duke Nicholas, of France's ally Russia, as well as by Sir Henry Wilson for the British. They set General Galliéni (Blue) against General Marion (Red), each force consisting of two army corps and a cavalry division. Galliéni secured victory on the final day with a spectacular charge by General Dubois' 1st Cavalry Division, taking the enemy's 9th Corps in the rear and capturing Marion and the corps commander along with their staffs, corps artillery and four aeroplanes. Ashmead-Bartlett of *The Daily Telegraph* argued that this incident disproved those who declared that modern warfare was becoming increasingly scientific thanks to the increasing role of railway schedules and aerial reconnaissance – 'Mistakes will still happen even in the most highly organised and scientific armies, … there is still scope for the individual brain of a commander to seize the psychological moment and change the fortunes of the day by a brilliant *coup de main*.' Ashmead-Bartlett professed himself impressed with what he saw of the French army in 1912 ('I had followed the operations of five Army Corps, and had seen them handled with machine-like precision, controlled, fed, and concentrated with such ease that war was made to appear a ridiculously easy game'), though this praise was designed to emphasise the contrast with the chaotic state of the Turkish army he witnessed fighting the Bulgarians in Thrace the following month.[34]

The Canadian contingent at the French manoeuvres made a number of observations which are of interest given that some of them attended the British ones immediately afterwards, thereby making direct comparison possible. They were critical of the French cavalry's reluctance to employ dismounted action, and noted the dense firing lines adopted by the infantry – 'Lines were thicker than in the British manoeuvres.' They did, though, concede that this might be due to the greater numbers of troops available to the French with an equivalent frontage. They also commented on the lack of a general service uniform, observing that 'consequently their clothing would appear

32 Macdiarmid, op. cit. p.219.
33 Ibid., pp.220-222.
34 E. Ashmead-Bartlett, *With the Turks in Thrace*, pp.1-2.

too conspicuous and distinctive for modern conditions.' The Canadians did, though, come away with the same impression as all visitors did of the French artillery, 'which has a gun that gives amazing rapidity of fire.'[35]

Henry Wilson, who was accompanied by Colonel George Macdonogh, found time during the manoeuvres to meet the Grand Duke Nicholas and the French Minister of War, Alexandre Millerand. The British visitors were treated with great hospitality by their hosts, finding themselves members of Joffre's mess, along with de Castelnau and four junior French officers. A convinced Francophile, Wilson was impressed with all that he saw except the cavalry, echoing the observations of the Canadian officers: 'They certainly are wonderful marchers. Cavalry men and horses excellent, but handling is all "*arme blanche*". So useless.' Wilson took away with him the following impressions: 'a) the ease with which these Frenchmen move, feed, and fight large masses of men; b) the marching of the infantry. The cavalry was *very* ill handled, as they would not dismount. Curious these Frenchmen being so obstinate about the "*arme blanche*."'[36] The War Office's official report conceded that the marching on show was superior to the British and expressed admiration for 'the enterprising and aggressive spirit animating all ranks'. However, the French were adjudged to be inferior to the British when it came to minor tactics – there was a lack of efficiency in fire direction or control and troops were too keen to close with the enemy: 'Like the cavalry, the infantry did not seem to realize what modern rifle fire is like.'[37] Remarkably, every country's army appears to have made the same criticisms of those they observed at foreign manoeuvres, chiefly that attacks were made in excessively dense formations, without sufficient allowance for the destructive impact of modern weaponry. Perhaps more to the point, every country's army found that visits to each other's manoeuvres reinforced deeply-held national stereotypes and confirmed them in their conviction that their own methods were the best.

As well as attending manoeuvres and exercises, there was one other way for British officers to see foreign armies in action, and that was, of course, to travel to a theatre of war. There were two significant conflicts in this period which afforded such an opportunity: the Russo-Japanese War (February 1904 – September 1905) and the Balkan Wars (October 1912 – May 1913, June-July 1913). Both wars were attended by numerous British observers: Nicholson, later to become Chief of the General Staff, was the official attaché to the Japanese Army while Lieutenant-General Sir Ian Hamilton, later to command at Gallipoli, represented the Indian Army in Manchuria. Aylmer Haldane, who would command a corps on the Western Front, also observed the fighting in Manchuria, and warned the 1909 General Staff conference of the

35 https://canadaatwarblog.wordpress.com/2015/06/07/canadian-officers-notes-on-french-army-manoeuvres-sept-11th-17th-1912/ (accessed 30/8/2016).

36 Maj. Gen. Sir C. Callwell, *Field-Marshal Sir Henry Wilson His Life and Diaries*, pp.116-117.

37 TNA WO 33/618, Report on Foreign Manoeuvres 1912, p.20.

impact of machine-guns on thick attacking lines.[38] Captain James Jardine of the 5th (Royal Irish) Lancers was a member of the British contingent sent to Tokyo in 1904, and when commanding a brigade at the Somme based his deployments on the first day of the battle on his observation of Japanese infantry tactics during the war.[39] Jardine also drew defensive lessons, contrasting the poorly constructed Russian trenches at Simaichi on 7 June 1904, which lacked head cover and an effective field of fire, with the well laid-out and concealed Japanese positions.[40] Captain Berkeley Vincent of the Royal Artillery was attached to the 2nd Division of the First Japanese Army, and on his return to Britain attended the Staff College, where he fell out with Henry Wilson, who rejected his assertion that superior Japanese morale had enabled the success of their massed infantry assaults on Russian defensive positions.[41]

Major Philip Howell had been one of those officers who had accompanied Thompson Capper on his battlefield tour of Manchuria in 1907; by now Senior Instructor at the Staff College and widely regarded as one of the most able officers in the army, Howell was sent to Thrace during the Balkan Wars and was attached to the Bulgarian Army of General Savov. On his return to England Howell briefed Douglas Haig on what he had seen (as did John Charteris) and his experiences would be turned into a series of lectures for the Staff College, and provide the basis for a book, *Campaign in Thrace – 1912*.[42] In light of his Balkan experience, Howell was sent to the Salonika theatre as Chief of Staff of the British contingent in October 1915 before returning to the Western Front in time for the Somme offensive, during which he was killed in action near Thiepval in October 1916.

The British produced three different multi-volume official accounts of the Russo-Japanese War, the last of them the *Official History of the Russo-Japanese War (Naval and Military)*, published in three large text volumes with two containers of maps. Sadly for those British officers wishing to learn lessons from the war, the final volume of the Official History was not completed until June 1914, and did not reach the publishers until after the First World War.[43]

Following the Russo-Japanese War there was much debate on the lessons to be learnt, especially from the Japanese victories at Port Arthur and Mukden, which had been achieved against strong Russian defences, though at great cost to the attackers. Many observers came to the conclusion that an army with high moral qualities and a spirit of patriotic self-sacrifice could overcome modern weapons. According to this interpretation, the effectiveness of offensive tactics on the modern battlefield had been

38 TNA WO 279/25, Report of a Conference of General Staff officers at the Royal Military College, January 1909, p.68.
39 M. Samuels, *Command or Control?: Command, Training and Tactics in the British and German Armies, 1888-1918*, p.144.
40 N. Murray, *The Rocky Road to the Great War: The Evolution of Trench Warfare to 1914*, p.150.
41 K. Jeffrey, op. cit., pp.82-3.
42 Brig. Gen. J. Charteris, *Field-Marshal Earl Haig*, p.67.
43 J. Sisemore, *The Russo-Japanese War, Lessons Not Learned*, p.110.

proven. Infantry assaults against prepared defences could succeed as long as they were carried out by troops with strong discipline and powerful national spirit.[44] The effect of the Russo-Japanese War was to cause a rethink in the British army.

The Boer War had apparently proved the dominance of fire power on the modern battlefield, but according to the Director of Staff Duties, Brigadier-General Kiggell, speaking at the 1910 annual General Staff conference, the war in Manchuria had proved this to be incorrect. 'Victory is now won actually by the bayonet, or by the fear of it.'[45] Major-General Altham went so far as to state that 'the Manchurian campaign has wiped out the mistaken inference from South African experiences that bayonet fighting belonged to the past.'[46] Tim Travers has argued that the period between 1902 and 1914 witnessed a regression from fire power to a belief in the spirit of the offensive, though it is important to recognise that Kiggell went on at the same conference to emphasise the need to combine fire with movement. British officers never went as far as their French counterparts in embracing the cult of the offensive, as seen in Major Rooke's sober assessment in a lecture to the RUSI in early 1914: 'It is clear that however much the attacking troops may be "trained above the fear of death" this itself will not prevent their being struck by the enemy's bullets, and may not improbably even increase their losses since such troops are likely to expose themselves unduly.'[47]

Although there was broad agreement about the part played in the Japanese victory by the offensive spirit, there was disagreement over details. While the French and Austro-Hungarians tended to discount the notion that anything could be learned from the fighting in Manchuria, the armies of Britain, Germany and the United States all made changes to their infantry doctrine after the war, incorporating extended order fighting, though the German Drill Manual of 1906 added the important caveat that that the breaking up of larger formations should be avoided whenever possible. The British Army's updated Field Service Regulations of 1909 established the use of forward rushes, combined with fire and manoeuvre, as a primary tactic.[48]

Another potential lesson from the war lay in the Japanese use of what might be termed an integrated weapons system, which included hand grenades, trench mortars and machine guns. Hand grenades, which the British had virtually phased out by 1905, proved especially effective in the close fighting at Mukden, although one British observer dismissed such reports as 'exaggerated' and insisted that there would be insufficient opportunities in a European war to justify training in the use of grenades.[49] Britain did introduce new types of grenades after the Russo-Japanese War, but only in very small numbers. In his memoirs, William Robertson claims that when DMT

44 Ibid.
45 T. Travers, 'The Offensive and the Problem of Innovation in British Military Thought 1870-1915', *Journal of Contemporary History,* p.531.
46 Maj. Gen. E. Altham, *The Principles of War,* p.205.
47 S. Jones, *From Boer War to World War*, pp. 65-66.
48 Sisemore, op. cit., p.111.
49 Ibid.

he asked for the troops to be supplied with dummy grenades, costing about twopence each, but was allowed penny grenades only, ' at a total cost of perhaps not more than thirty or forty pounds', while his request for live grenades for demonstration purposes was turned down altogether (they cost nearly £1 each).[50]

As for trench mortars, nothing was done to develop one until after the outbreak of war, when the BEF resorted to using improvised wooden mortars. The Germans did more, developing small mortars to engage entrenched defenders, but even they only possessed 160 mortars at the start of the war.[51] By 1917, the Allies were using seven different types of trench mortars on the Western Front and Britain was producing 4,000 hand grenades a month by March 1915.[52] Finally, while the effectiveness of machine guns was clearly demonstrated, both in attack and defence, foreign armies were slow to increase the number of machine guns they allocated to each division. This was partly due to considerations of cost and partly due to a long-running disagreement over their use. Was the machine gun primarily a defensive weapon to be massed in the firing line, or a 'weapon of opportunity', to be kept on the flanks with the cavalry? This confusion is the source of the oft-misinterpreted words of a colonel at the 1910 manoeuvres, who when asked by a subaltern what should be done with the machine guns responded: 'Can't you see I'm busy? Take the damn things to a flank and hide them.'[53]

The lessons of the Russo-Japanese War, which were only partially learned by the British Army, were reinforced by the two Balkan Wars. In the first (October 1912-May 1913), the Balkan League (Greece, Bulgaria, Serbia and Montenegro) defeated the Ottoman Empire and annexed large amounts of Turkish land. The second (June-July 1913) saw the former allies falling out over the spoils of victory, with the Bulgarians fighting Greece, Montenegro, Serbia and Romania, while the Ottoman Empire also joined in to recover land lost earlier. All six of the participants would take part in the First World War, which started in the region a year later, and it could plausibly be argued that what we call the First World War was in reality the Third Balkan War, at least in its origins.

It might seem surprising that so little was learnt from the two Balkan wars, given that they featured many of the characteristics of the First World War and took place on several of the battlefields of the later conflict, including Salonika, Gallipoli and the Dardanelles. The fighting between the Bulgarians and the Turks at Chadaldzha (November 1912) bogged down into protracted trench warfare, and heavy casualties were inflicted on massed infantry assaults by modern rapid-firing guns.

There are a number of reasons why lessons from the Balkan Wars were not learnt by the armies of the Great Powers, Britain included. There was a tendency to assume that

50 Field-Marshal Sir W. Robertson, *From Private to Field-Marshal*, p.192.
51 A. Saunders, *Trench Warfare 1850 – 1950*, p.135.
52 A. Rawson, *The British Army 1914 – 1918*, p.262.
53 R. Holmes, *Riding the Retreat: Mons to the Marne 1914 Revisited*, p.58.

Machine guns in use at the Irish Command manoeuvres c.1911. (Mary Evans Picture Library)

little could be learnt from armies made up of peasant conscripts in an under-developed part of Europe such as the Balkans, though Howell was impressed by the Bulgarian army's fighting qualities. There was also the problem that the Balkan armies were reluctant to allow journalists or military attachés from the Great Powers to have access to the combat zone. Those that did reach the front line seemed reluctant to learn – the American Lieutenant Sherman Miles observed of his European colleagues that 'The attachés were extremely keen on the strategy of the campaign… but the construction and position of the trenches, the effect of fire on the trenches, the lay of the country and the handling of men, they did not regard as of much interest to General Staff officers.'[54] However, the main reason that the lessons of the Balkan Wars made little impression on British, French and German military thinking is that the First World War simply followed too quickly for there to be time for those lessons to have been absorbed.

In any case, there *were* sincere attempts to learn from events in the Balkans. There was widespread international interest in the use of air power; the Bulgarians, Romanians, Greeks and Serbs all employed aeroplanes for reconnaissance purposes, and the Bulgarians also used aeroplanes to bomb Adrianople (30 November 1912), the first time bombs had been dropped on a European city. Articles on the fighting

54 R. Hall, *The Balkan Wars 1912–1913: Prelude to the First World War*, p.134.

Bulgarian artillery during the First Balkan War, 1912. (Alamy)

appeared in the *Cavalry Journal* and the *Journal of the Royal United Services Institution* and there was an intense debate in the French and German armies surrounding the role of the artillery in the Turkish defeat; typically, both sides of the argument interpreted the results of the conflict as vindicating their own equipment and tactics. The victories enjoyed by the armies of the Balkan League, which were equipped with French Schneider 75mm medium guns, was argued by the French to have proved the superiority of their artillery, while the German General Heinrich von Rohne argued that the Turkish success at Chadaldzha, where well-prepared fortifications and howitzers and heavy artillery had stopped the Bulgarians in their tracks, demonstrated the need for longer range and heavier artillery for the coming conflict.

Observers on all sides agreed that the massed infantry assaults of the Bulgarians and Serbs confirmed the lesson of the Russo-Japanese War, that Japanese-style attacks in close order could succeed, though at considerable cost, as long as they were pressed with sufficient determination.[55] One British observer wrote of the Bulgarian infantry 'absolutely throwing away their lives in the Japanese manner whenever a point had to be taken or won'.[56] Thus the main effect of the Balkan Wars was to reinforce existing patterns of military thinking, with any contrary evidence conveniently ignored, another case of cognitive dissonance. However, one or two individuals did draw the correct conclusions from a study of the Balkan War – the British writer Captain A.H. Trapmann, who had accompanied the Greek Army in its assault on Janina, noted:

55 Hall, op. cit., p.134.
56 Ashmead-Bartlett, op. cit., p.160.

'The machine gun is obviously *the* weapon of the future; in the opinion of this author the next war between first-class military powers will establish the vast superiority of the machine gunner over the rifleman.'[57] Trapmann also foresaw an important role for hand grenades in the coming conflict.

The Balkan Wars highlighted another problem which would be encountered in the First World War, even if few heeded the warning, namely the massively increased consumption of ammunition by the new rapid-firing guns. In this respect it was confirming the experience of army manoeuvres, whether French, German or British. During the Russo-Japanese War, the Russians' average expenditure of artillery ammunition was 87,000 rounds a month, scarcely changed from that of the Germans in the Franco-Prussian War (81,000 a month). In contrast, the Bulgarians' average monthly expenditure of artillery ammunition in 1913 was 254,000 a month, an utterly unforeseen rate of increase which excited comment in military journals, one of which reported that Bulgaria had decided to increase by 50% its war reserve stocks of artillery shells.[58] Yet none of the Great Powers responded in the same way and each entered the war in August 1914 trusting that its pre-war stock would suffice. All the combatants would be plagued by shell shortages in the early years of the war – indeed, the British government would fall thanks to the issue in May 1915. This was one lesson from the Balkan Wars which was clearly not heeded.

57 Captain A.H. Trapmann, *The Greeks Triumphant*, p.294.
58 D. Zabecki, 'The Dress Rehearsal: Lost Artillery Lessons of the 1912-1913 Balkan Wars', *Field Artillery*, p.22.

7

## 1914: The B.E.F. goes to war

On 5 June 1914 General Grierson wrote to his relative, the Rev. Robert Grierson to inform him of his plans for the coming months:

> In September I am going to command one side at our Army Manoeuvres and fight my old enemy (and close friend) Douglas Haig. The manoeuvres are to be in the Severn Valley and all the regulars in England are to take part in them. I only hope that there will be no Irish troubles or strike duty to interfere with them.[1]

In the event the 'troubles' which ensured that these manoeuvres never did take place were not domestic but took place far away, in the Balkans. The assassination of the Archduke Franz Ferdinand in Sarajevo on 28 June and the ensuing July Crisis initiated a process which would bring about Britain's entry into a war alongside Russia and France against Germany and Austria-Hungary. Following the British declaration of war on Germany on 4 August, the value of all the training and exercises the British Army had been engaged in for the previous twelve years would be tested as the BEF found itself in the path of the German advance through Belgium.

The War Council held on the afternoon of Wednesday 5 August brought cabinet ministers and senior commanders together, including the newly-appointed Secretary of State for War, Field Marshal Lord Kitchener, for a discussion which provided a dispiriting foretaste of the problems the generals would have with the politicians over the next four years; Wilson described them 'discussing strategy like idiots', adding that 'Jimmy Grierson spoke up for decisive numbers at the decisive point. Sir John French said we should go over at once and decide destination later.'[2] French later wrote that those present felt an obligation to send to France as strong as an army as they could, and 'there was an idea that one cavalry division and six divisions of all arms had

1 D. Macdiarmid, *The Life of Lieut. General Sir James Moncrieff Grierson*, pp.253-254.
2 Maj. Gen. Sir C. Callwell, *Field-Marshal Sir Henry Wilson, His Life and Diaries*, p.102.

been promised.'[3] This meant an expeditionary force of 100,000 men. However, after all the pre-war planning and the assurances that the Royal Navy's command of the Channel guaranteed that Britain would not be invaded, it was decided that only four divisions, along with the Cavalry Division, would be sent straight away. Two divisions would be kept behind in Britain to guard against a German invasion, though in the event both would soon be sent to France. The promise made by Henry Wilson to the French, which Kitchener only learnt about at the War Council, had been broken, or at least diluted. It was not a promising start.

The decision was also taken that the BEF would concentrate at Maubeuge as originally planned rather than further back at Amiens as Kitchener advocated or further forward at Antwerp as the mercurial French suggested, a proposal which greatly annoyed Haig, Wilson and Grierson. British mobilisation proceeded with remarkable efficiency, thanks to the planning which had been carried out over the previous four years. The War Book, first issued in 1912, and War Office files such as WO106/49B/1 (Instructions for Entrainment of the British Expeditionary Force) ensured the efficient mobilisation of troops, recall of reservists and despatch of units ready for the BEF to start to cross to France on 9 August, heading for Boulogne, Le Havre and Rouen.[4]

Sir John French arrived at Boulogne on 14 August along with his Chief of Staff, Sir Archibald Murray, and Sir Henry Wilson, apparently French's own choice for the job and now compensated with the nebulous post of Sub-Chief of Staff, an appointment which could only serve to undermine Murray. As adumbrated in the 1912 manoeuvres, Haig would command I Army and Grierson II Army (promptly renamed Corps to fit in with their French allies), with Allenby commanding the Cavalry Division. The two Army commanders crossed to Le Havre with their staff on 15 August on SS *Comrie Castle*. The opportunity for Grierson to prove in action the potential hinted at in his 'victory' over Haig in 1912 appeared to have arrived, but it was not to be. Never one to stint himself at the dinner table, by 1914 his girth had expanded to alarming proportions, and he recorded, ominously, that he had started 'a course of rubbing for swelling in my hand (?gout)'.[5] He was later described as 'looking like a beef extract advertisement' when he left Southampton.[6] At about 7:30 a.m. on 17 August, in a train carriage en route to Amiens, Grierson died of an aneurism of the heart, attended by his ADC 'Cakes' Banbury (who would be killed at the Battle of the Aisne within a month). Captain Dunn recorded the circumstances:

3 Field Marshal Viscount French of Ypres, *1914*, pp.6-7.
4 Robin Neillands, *The Old Contemptibles, The British Expeditionary Force, 1914*, p.90. Neillands provides an excellent and detailed analysis of the logistics behind the BEF in 1914.
5 D. Macdiarmid, *The Life of Lieut. General Sir James Moncrieff Grierson*, p.255.
6 Major-General J.F.C. Fuller to Sir Basil Liddell Hart, 6 May 1937.

> I was summoned early to LoC HQ and told that General Grierson, commanding II Corps, had died suddenly in the train… … There were a lot of stupid rumours about General Grierson's death. I have the best of reasons for knowing that he died from the bursting of a blood-vessel, probably brought on by the heat and a heavy meal. He was a man of full habit, the weather was torrid, and the Staffs of higher formations were at that time living exclusively on hampers supplied by Fortnum and Mason.[7]

At about 9:00 a.m. Haig, travelling in the train immediately in front of Grierson's, was summoned to the telephone at the wayside station of Serqueux and informed by the station master at Morgny of his old friend's death. With characteristic decisiveness and lack of sentimentality, Haig gave orders for his train to continue to Amiens, reasoning that 'It seemed to me of no use to delay… … as I could do nothing.'[8] Haig's Intelligence Officer, John Charteris, had noted at Le Havre that Grierson was 'feeling the strain and complains of not feeling well' and now lamented that 'poor Grierson was of too full a habit to stand even the comparatively comfortable strain of soldiering at Corps HQ.'[9] The impact of Grierson's death was felt far beyond Amiens. Captain Roly Grimshaw of the Poona Horse, on the troop ship SS *Dongola* between Gibraltar and Malta, recorded in his diary for 18 August: 'So Grierson is dead. He looked apoplectic.'[10] Grimshaw had been on leave in England when war broke out, so he may well have been speaking from personal observation. On 6 September, Alexander Scrimgeour, a nineteen-year-old midshipman writing to his mother from the cruiser HMS *Crescent* in the North Sea, reported that 'there is a story being circulated that General Grierson, who is supposed to have died a natural death, was murdered by a German because of his extensive knowledge of the German Army and its methods.'[11] This must have been one of the 'stupid rumours' referred to by Captain Dunn, and is testimony not only to the fevered spy scare gripping Britain at the outbreak of war but also to Grierson's renown as an expert on the enemy.

The death of one of the two corps commanders was a severe blow to the BEF before a shot had been fired, but the damage was exacerbated by Lord Kitchener's choice of Grierson's successor. The vastly experienced General Sir Ian Hamilton offered his services, but French requested that Lieutenant-General Herbert Plumer be sent out to take over II Corps. However, Kitchener insisted on the selection of Smith-Dorrien, despite French's animosity towards Sir Horace, which was well known to all senior officers in the army. It is likely that Kitchener wanted someone who would stand up to French, but the choice, and the tension it inevitably created in the high command

7 Captain J. Dunn, *The War the Infantry Knew 1914-1919*, p.14.
8 Lt. Gen D. Haig (ed. G. Sheffield and J. Bourne), *War Diaries and Letters 1914-1918*, p.59.
9 Brig. Gen. J. Charteris, *At GHQ*, pp.10-12.
10 Captain R. Grimshaw, *Indian Cavalry Officer*, p. 17.
11 A. Scrimgeour, *Scrimgeour's Small Scribbling Diary 1914-1916*, p. 50.

Allenby, Grierson and Haig sitting with the French Commandant at Mailly, 6 July 1914. (John Mackenzie @Britishbattles.com)

of the BEF, would have serious consequences in the coming weeks and months, above all at Le Cateau.

It is, of course, intriguing to speculate how different things might have been had Grierson, and not Smith-Dorrien, commanded II Corps in August and if he had been available as an alternative to Haig as French's replacement in December 1915, but anyone glibly proposing Grierson as the BEF's 'lost leader' must confront the issue of his physical unfitness – if he could not cope with a train journey to the front, how could he have withstood the strain of the retreat from Mons? This piece of counterfactual history therefore falls at the first hurdle. Mark Connelly has also drawn attention to the fact that Grierson exhibited the same limitations as many of his colleagues when it came to tactics.[12] Grierson was a product of his training and the prevailing military culture in the British Army, and it is likely that he would have led II Corps in a similar, if less short-tempered, fashion to Smith-Dorrien, with whom he had enjoyed a cordial relationship at Aldershot.[13] The qualities which made Grierson such

12 M. Connelly, 'Lieutenant-General Sir James Grierson' in S. Jones (ed.), *Stemming the Tide*, p.145.
13 Ibid., p.149.

a successful general on manoeuvres are likely to have made him an effective corps commander in the mobile warfare of the first month of the war, but there is no guarantee that they would have made him a better commander of the BEF than Haig in the long run. The scale and complexity of the fighting on the Western Front called for a different style of leadership from the heroic, personal model which was in vogue before 1914 and which better suited a colonial army, and the general on the British side who most closely fitted the bill was not Jimmy Grierson but Douglas Haig. Interestingly, Liddell Hart was of the opinion that Grierson, with his knowledge of the Germans and excellent relations with the French, would have made a very suitable Chief of Staff in place of the inadequate Murray.[14]

A number of important lessons could be drawn from the pre-war manoeuvres, many of which could now be put into practice as the BEF came into contact with the German army. The RFC's first reconnaissance flight of the war was made on 19 August 1914, when Captain Joubert de la Ferté flew from Maubeuge in his Blériot XI; this was precisely eleven months after he had flown the same aeroplane in the skies over Buckinghamshire as part of the Brown forces in the 1913 exercise. The potential of aerial reconnaissance had been amply proven in 1912 and 1913, and would be again at two pivotal moments in the opening weeks of the war. On 22 August a reconnaissance flight by Lieutenant Vivian Wadham (another pilot in the Brown air force of 1913) and his observer, Captain Lionel Charlton in another Blériot XI informed the BEF of the German advance towards them when a column of German infantry from von Kluck's First Army was spotted; this information helped to decide French against continuing the BEF's advance, paving the way for the defensive action at Mons the next day. It is significant, however, that French needed further confirmation, including from the cavalry, before he would believe these reports. It seems that memories of Haig's difficulties at the 1912 manoeuvres still played on French's mind.

The second key moment came on 3 September when aerial reconnaissance spotted I Army's change of direction to the east of Paris and reported the resultant gap between Kluck and Bulow's II Army. This information helped to enable the Allied counter-attack at the Battle of the Marne and ensured the end of German progress towards Paris.

The importance of camouflage and concealment had also been demonstrated; commanders were now well aware of the need to conceal their troops' movements from aerial spotting where practicable. On the retreat from Mons, troops dispensed with the wire that made their caps flat and rigid, thereby making them less uniform and easily spotted from the air, and they soon learnt to remove their cap badges which reflected the sun, to be replaced much later by a blue cloth alternative.[15] Much later in the war, Allenby made masterly use of camouflage when massing his cavalry at

14 B. Liddell Hart, *History of the First World War*, p.33.

15 Ben Clouting, *Teenage Tommy – Memoirs of a Cavalryman in the First World War*, edited by Richard van Emden, pp. 60-61.

Megiddo in September 1918 – he hid his cavalry in orange groves just north of Jaffa, and dummy wooden horses were covered with blankets and set out in lines for German planes to photograph.[16] Perhaps the lessons taught in September 1912 had been learnt by Allenby, who had been on the receiving end of Grierson's successful concealment of Snow's division. Mistakes could still be made, as had been shown in 1912; during the first Battle of Ypres in October 1914 observers from No. 6 Squadron, a unit which had not been involved in the 1912 or 1913 manoeuvres, mistook the shadows cast by gravestones for a bivouac in a churchyard, while long patches of tar on a road were taken to be troops on the march.[17]

Another important element in reconnaissance, at least at the outset of the war, was the contribution made by cyclists, who were adjudged to have proved their worth in the manoeuvres held before the war; Bennet Burleigh of *The Daily Telegraph* observed after the 1912 manoeuvres that 'the cyclist is a coming type in war'.[18] Whereas Allenby had been reluctant to employ the Red cyclists to support the Cavalry Division, the Blue cyclists had carried out a variety of roles, scouting and occupying positions as well as protecting the flanks of infantry units. Following the 1912 manoeuvres it was agreed by the Army that cyclists had an important role to play in combination with cavalry; as well as their scouting role, they could offer fire support and free the cavalry for mounted action, as the Essex Cyclists had done to great effect during the 1913 exercise.[19] Cyclists were mobile, quiet and cheap and their bicycles were easily transported and did not need fodder or water as horses did, though they did depend on good roads and were vulnerable without cavalry protection.

In fact, Burleigh's forecast proved to be wide of the mark. The BEF, like its continental counterparts, contained large numbers of cyclists in August 1914 – John Parr, the man thought to have been the first British soldier killed in the Great War, was a cycle scout with the Middlesex Regiment attached to 3rd Division. In the early stages of the war the cycle companies of 100 men attached to each division so proved their worth when working with cavalry that the War Office decided to double their size. However, the development of trench warfare proved as detrimental to the cyclists as it did to the cavalry. The Army's fourteen Cycle Battalions were mostly deployed in Britain or used to replace infantry losses, and it was not until the return of conditions of mobile warfare in the last months of the war that cyclists could again prove their value as they had in 1912.

The Territorial infantry and Yeomanry cavalry had exceeded expectations in both 1912 and 1913 – the South-West Mounted Division, made up entirely of Yeomanry, had performed especially well against regular cavalry in 1912. Grierson spoke highly

16 Maj. Gen. J. Vaughan, *Cavalry and Sporting Memories,* p.9.
17 Sir Walter Raleigh, *The History of the War in the Air 1914-1918*, p.304.
18 *The Cambridge Daily News*, 21 September 1912.
19 TNA WO 279/48, Report of a Conference of General Staff Officers at the Royal Military College 1913, p.40.

Cyclists in action at Shrivenham during the 1909 manoeuvres. (Author's collection)

of the Territorials' performance, though he did suggest that they needed a greater allocation of staff officers 'as they have to be "nursed" considerably more than Regulars'.[20] Smith-Dorrien observed that the Territorials presented 'considerable possibilities' if suddenly needed in war time, though insisting that they had all the weaknesses inevitable in troops 'very imperfectly trained, and lacking battle discipline'. He was especially critical of their lack of perception when an opportunity presented itself, as in the incident on 17 September 1912 when half a company of Territorials watched with indifference a squadron of Red Cavalry crossing a bridge at a walk, in close formation, only 500 yards from their trenches. He criticised the lack of cohesion in their attack on the afternoon of 18 September – 'a great deal too much noise and shouting' – but also commended the 'endurance, cheerfulness and soldierly spirit' of the 7th battalion of the Liverpool Regiment, posted on the extreme left of the outpost line. He noted that they led the attack that day having already marched over 20 miles to get there. In a battalion 200 of whom had enlisted since May, not a single man fell out on the line of march and the attack was carried out 'with the energy and keenness of troops fresh from barracks'.[21] Their performance in September 1912 was a precursor to the Liverpool Regiment's significant contribution when war came; all four battalions were among the first Territorial units to be committed to France, and all four would be commanded by the men who had led them in September 1912.

20 TNA WO 279/47 Report on Army Manoeuvres 1912, p.173.
21 WO 279 / 47, Report on Army Manoeuvres, 1912, p.140.

Men of the 7th King's Liverpool Regiment, probably taken at a training camp. (Courtesy of Ronnie Cusworth at the Litherland and Ford website)

Even so, the arrival in France in March 1915 of the 1/7th Battalion King's Liverpool Regiment was greeted with some trepidation by Regular commanders, who were sceptical of the Territorials' ability to take on front line duties. In a document circulated to First Army Divisions in February 1915, Robertson was critical of the Territorial units arriving in France, identifying rifle and machine gun training, march discipline, entrenching, communications and platoon leading as areas in need of urgent attention, but this criticism was aimed at the quality of their training in England in the seven months since the war started rather than their pre-war training.[22] The 1/7th Liverpool Regiment were held to have performed well in their first engagement, at Festubert in May 1915, but the men of the battalion were well aware of Regular attitudes towards them and were keen to make a good impression; one member of the 1/7th Liverpool Regiment wrote after Festubert: 'We have covered ourselves with glory, and have gained great respect from the Regulars', while another, Private Pirrie, reported officers shouting to their men, 'We are only Territorials but we'll show them what we are made of.'[23]

22 A. Gregson, "Seven Days in Hell" The 1/7th Battalion King's Liverpool Regiment at the Battle of Festubert, May 1915 in S. Jones ed. *Courage Without Glory: The British Army on the Western Front 1915*, pp.286-287.
23 Ibid., p.307.

The positive impression made by the Territorials in 1912 and 1913 is confirmed by the remarks of an influential French observer, General Hippolyte Langlois, who observed Territorial battalions engaged with Regulars in divisional manoeuvres in August 1909. Langlois was a former Commandant of the École de Guerre who had commanded XX Corps between 1901 and 1904 and had been a member of the Conseil Supérieur de la Guerre; he was part of a French delegation invited by the London press which included Ludovic Naudeau, who had been taken prisoner by the Japanese while reporting on the war in Manchuria, and later reported on the Balkan Wars, and Reginald Kann, who had seen action – on the Boer side – in South Africa, as well as reporting on wars in Cuba, Manchuria and Morocco. Langlois witnessed Territorials from the London Regiment training on Salisbury Plain with Regulars, including the Scots Greys, and later travelled to Flintshire where he watched the West Lancashire Division on manoeuvres. He professed himself impressed by the keenness of the men, the food and supply arrangements and the medical services, and concluded that 'this Territorial Army, at any rate its infantry, is not a worthless "national guard", but a militia which even now is a factor to be reckoned with.'[24]

Langlois went on to make the interesting observation that in one respect the Territorials were superior to the Regulars. He argued that Regular soldiers were men attracted by the pay, whereas the Territorial Army 'is composed of volunteers who serve from motives of duty and patriotism'. The Territorials might not be able to match the Regulars in terms of technical proficiency, 'but its moral force is greater', a far from negligible factor at a time when offensive spirit and moral forces were considered to be of such importance in the aftermath of the Russo-Japanese War. It is likely that Langlois' judgement was affected by a traditional French preference for a citizen army, but elsewhere he was complimentary about the Regulars, and at the very least he can be taken as an advocate of the Territorial force only a year after its creation.[25]

However, though foreign observers might be impressed, and other commanders were forced to revise their opinion of the capabilities of Territorial units, Lord Kitchener never did change his view of them.[26] When Kitchener decided in August 1914 to raise 70 new divisions over the next three years, he chose to do so outside the existing Territorial system. Kitchener, who had spent most of his career in India or Africa, had a much lower opinion of the value of the Territorials than French, Haig, Smith-Dorrien or Rawlinson, all of whom had seen them in action in pre-war manoeuvres. His low opinion of part-time soldiers was in large part due to his experience of French territorial soldiers, a very different beast from the British version. [27] Haig expressed shock at Kitchener's 'ignorance of the progress made by the Territorial

24 General H. Langlois (translated by Captain C.F. Atkinson), *The British Army in a European War*, p.22, p.36.
25 Ibid., pp.38-39.
26 I. Beckett, 'The Territorial Force' in I.F.W. Beckett and K Simpson (eds.) *A Nation in Arms: A Social Study of the British Army in the First World War*, p.131.
27 K.W. Mitchinson, *The Territorial Force at War, 1914-1916*, p.39.

Army towards efficiency. Personally, I was very intimately acquainted, of course, with what the Territorials had been doing.'[28] Haig appears to have had an open mind when it came to the Territorials; while DMT back in March 1908, he had written to Richard Haldane to express his view, based on his positive impression of the Lanarkshire Artillery Volunteers at the previous year's Scottish manoeuvres, that Lord Roberts was wrong to argue that field artillery should be omitted from the scheme for the Territorial Army.[29]

Manoeuvres had provided invaluable experience in terms of supplying and transporting large numbers of troops. In 1909 the Great Western Railway conveyed approximately 514 officers, 14,552 men, 208 officers' horses, 2,474 troop horses, 25 guns, 34 limbers and 581 wagons and carts in the manoeuvres of that year – 'The military authorities and the Army contractors expressed their pleasure at the manner in which the work was performed by the Company's staff.'[30] In 1912 Major F.G. Fuller of the Royal Engineers noted that the London & North Western and Great Northern Railways were able to transport to East Anglia and detrain the 16,348 men and 3,168 horses and 832 guns and vehicles of the Blue infantry divisions 'with extraordinary punctuality and without a single hitch of any consequence'. Nearly 200 troop trains (4,000 vehicles) were employed in 1912, of which 50% started exactly on or even before time, with the rest only a few minutes late. One valuable lesson learnt was that that if troops and supplies were off-loaded over the sides of carriage trucks with drop sides, as 4th Division were, it took half the time that it did if they were off-loaded from an end-loading dock as was the case with 3rd Division (18 ½ minutes for each train as opposed to 33 minutes). This was precisely the sort of problem-solving exercise which justified the Army's use of large-scale manoeuvres – the local press had been right to observe that the 1912 manoeuvres 'are in some measure a test of the efficiency of our Army and our railways'.[31] Another more prosaic but still significant observation was made by Smith-Dorrien who argued that longer halts were needed for the men to answer the call of nature: 'Unless men are given opportunities to urinate at scheduled halting places, they are bound to do so out of carriage windows, a practice which though of no importance in active service, is unseemly at manoeuvres.'[32]

Before 1912 Britain still lagged behind the continental countries in creating an administrative framework by which the State could exercise control over the railway system, and it was not until two months after the East Anglia manoeuvres that the Government created the Railways Executive Committee, containing representatives of all the major rail companies, to ensure control and co-ordination of the rail network in the event of war; the fruits of this decision would be seen in the considerable

28 D. Haig, *War Diaries and Letters 1914-1918*, p.55.
29 D. Cooper, *Haig*, pp.110-112.
30 E. Pratt, *Rise of Rail-power in War and Conquest 1833-1914*, p.196.
31 *The Cambridge Daily News*, 16 September 1912.
32 TNA WO 279/47, Report on Army Manoeuvres 1912, p.142.

administrative achievement of the BEF's deployment to France in August 1914.[33] As Hugh Drummond, the Chairman of the London and South Western Railway Company told a meeting of shareholders in February 1915, 'I think I may say that if the country was not prepared and equipped for war, the railways of the country were, at any rate.'[34]

A further lesson taught by manoeuvres was that greatly increased numbers of artillery shells would be needed in the coming war. After the experience of the Boer War and observations from Manchuria, many observers had warned about the vast amounts of ammunition the new rapid firing guns would expend. After witnessing the Russians use 120,000 rounds, weighing approximately 800 tonnes, in four days at Liaoyang in August 1904, Colonel Frederick Wing warned that 'the power of expenditure far exceeds the power of supply, the limit of the latter fixes that of our gun power.'[35] This was graphically confirmed in September 1912: Brigadier-General E. Phipps-Hornby VC, CB, the umpire attached to 4th Division's Artillery, compiled statistics on the amount of ammunition expended by each brigade on 18 September, which showed that battery commanders, firing in accordance with what they and their brigade commanders believed to be the tactical requirements, managed to expend in one day's battle between 70% and 95% of their entire ammunition in the field (for example, 32nd Field Artillery Brigade fired 2160 rounds of blank ammunition that day).[36]

The Germans came to the same conclusion following an experiment at their 1913 manoeuvres; firing under simulated battle conditions for 27 hours, a regiment exhausted 11,633 shells, an average of 323 shells per gun. As a result, Prussian War Minister Erich von Falkenhayn agreed to a target of 1,200 to 1,500 shells per gun, but by 1914 German field guns (987) and howitzers (973) still had well under the target and were some way behind the French. This confirmed what General Schubert had warned the Kaiser in December 1910: 'We will have shot up everything after the first victorious battles, and it will take months to replace it.'[37]

This lesson would be reinforced by the Balkan Wars, which witnessed an enormous increase in the volume of shells consumed by the artillery. Even if the lessons of pre-war manoeuvres and the Balkan Wars had been heeded by the Great Powers it was one thing to recognise the problem and quite another to remedy it; when war came it soon became clear that Britain's munitions industry was totally incapable of meeting the BEF's shell requirements, and in this it was not alone, despite the

33 Pratt, op. cit., pp.195-197.
34 E. Pratt: 'The Role of Railways in the War' in *The Great War, The Standard History of the All-Europe Conflict* Volume 6 p.143.
35 T. Bowman and M. Connelly, *The Edwardian Army*, p.81. Wing was involved on the staff in the 1912 Manoeuvres and commanded 3rd Division's artillery in August 1914. He was killed at Loos in October 1915.
36 Phipps-Hornby would command the artillery of III Corps in the BEF in August 1914.
37 E. Dorn Brose, *The Kaiser's Army: The Politics of Military technology in Germany during the Machine Age, 1870-1918*, p.151.

The Royal Field Artillery on the move, 10 September 1912. (Saanich Archives)

widespread belief that the Germans enjoyed a massive superiority – Lieutenant-General Pulteney reported on 16 September 1914 that 'the expenditure of ammunition is incredible you would not believe it possible for the Germans to have had so much available they must have been manufacturing for years while our ordnance people have been economising.'[38] In fact, Oliver Lyttelton, later Lord Chandos, a director of Metallgesellschaft between the wars, described how despite the supposed thoroughness of German preparations the firm's directors had been summoned by the General Staff before the end of 1914 and told that the Army was running out of copper and drastic measures were required, including the requisitioning of taps, door handles and church bells.[39] French planners entered the war confident that production of 3,600 shells per day as stipulated by their mobilisation plans would be more than sufficient. By 10 September 1914, the French Army had already used up over two thirds of its artillery ammunition, and by the end of the year it was consuming 900,000 rounds a month (the Germans would expend eight million rounds a month by 1918).[40] As one British gunner observed after the war: 'You cannot *devastate* if you are counting shell all the time.'[41]

38 A. Leask, *Putty – From Tel-el-Kebir to Cambrai,* p.205. The lack of punctuation is as in Pulteney's letter to Lady Desborough.
39 John Hussey, 'Without an Army, and Without any Preparation to Equip One', The British Army Review, p.82.
40 Major D. Zabecki, 'The Dress Rehearsal: Lost Artillery Lessons of the 1912-1913 Balkan Wars', *Field Artillery*, pp.18-23.
41 D. Clarke, *World War I Artillery Tactics*, p.19.

Cavalry doctrine had been dominated by a polarised debate raging since the Boer War which had pitted advocates of the *arme blanche*, notably French and Haig, against those such as Roberts who argued that the lance and the cavalry charge no longer had a place on the modern battlefield. The course of this debate was typified by the issue of the lance, which was dropped for all but ceremonial duties in 1903, only to be reintroduced in 1909. By the time of the 1912 Manoeuvres, the debate appeared to have been resolved by a doctrinal compromise – *Cavalry Training 1912* stated that 'Cavalry must… keep in close touch with the other arms and take advantage of their progress, offering them as much help as it can with its guns and rifles, and supplementing their decisive attack by sweeping the battlefield with its squadrons.' This did not rule out a role for massed formations, which could still 'achieve great results against troops which are worn out, demoralised, and obliged to leave their fighting positions'.[42] According to Spiers, the cavalry on the eve of war spent 80% of their time on shock tactics, 10 per cent on fire tactics, and 10 per cent on reconnaissance, figures which might suggest that the reformers had made little headway.[43] Haig was still thinking in terms of massed cavalry action in 1909: 'Occasions for charging will be few, but they occur – and the results from such actions will be immense. The mounted attack, therefore, must always be our ideal, our final objective.'[44] However, in many ways the cavalry were the British Army's most tactically sophisticated arm on the eve of war, and even though Douglas as Inspector-General still complained that 'our cavalry commanders are inclined to employ shock action whenever possible regardless of circumstances',[45] events in 1914 suggest that the cavalry, at least, had learnt from their pre-war experiences.

At Cerizy (28 August), 5th Cavalry Brigade, led by Brigadier-General Chetwode (the Cavalry Division Umpire in 1912), carried out a textbook combined fire and charge action. While A and B Squadrons of the 12th Lancers were sent on a wide outflanking movement, engaging the enemy dismounted, C Squadron carried out a successful frontal charge against dismounted German cavalry in combination with the Scots Greys. Fire support was provided by J Battery, Royal Horse Artillery and two machine gun sections while the 20th Hussars turned the enemy's other flank. Shrapnel from J Battery's four 13 pounders was especially effective in dealing with the infantry of the Guard Rifles, who had been forming for a counter-attack.

At Nery three days later, Briggs' 1st Cavalry Brigade employed superior firepower to defeat an attack by superior German forces; Brigadier-General Vaughan praised the 5th Dragoon Guards for having executed 'one of the best-timed cavalry rifle attacks' that he had ever experienced.[46] At Mont des Cats (12 October), the 16th Lancers

42 S. Badsey, *Doctrine and Reform in the British Cavalry 1880-1918*, p.240.
43 E. Spiers, 'The British Cavalry 1902-1914', *Journal of the Society for Army Historical Research*, p.79.
44 Badsey, op. cit. p.198.
45 TNA WO27/508, Report of the Inspector-General 1912.
46 Vaughan, op. cit., p.160.

advanced at the gallop, dismounted and opened rapid fire, after which covering fire from their machine guns helped them to capture the German trenches. The 4th Hussars guarded against any counter attacks from the east while the 5th Lancers were held in reserve and D Battery of the Royal Horse Artillery offered fire support at a range of 800 yards.[47] Once again, the British cavalry had carried out a model of flexible mounted tactics which belies the supposed resistance to innovation of their commanders.

Even the costly charge by the 9th Lancers and 4th Dragoon Guards at Audregnies (24 August) featured fire support from dismounted 18th Hussars on one flank along with the RHA.[48] These tactics had been honed in pre-war manoeuvres; almost all of the units had taken part in the 1912 manoeuvres, when Briggs' Blue cavalry had carried out a number of similar operations combining fire and shock, and all eleven of the Regular cavalry regiments involved in 1912 crossed to France with the BEF. Pre-war manoeuvres also taught the cavalry the need to act in concert with infantry. The 1910 manoeuvres witnessed a brilliant example of a combined operation between cavalry and infantry, when Brigadier-General Hew Fanshawe used his Cavalry Division in combination with the Oxford and Buckinghamshire Light Infantry, which had been placed at his disposal by Brigadier-General Edward Montagu-Stuart-Wortley, to fall upon the enemy cavalry, inflicting heavy losses.[49]

During the 1913 Irish divisional manoeuvres, Pulteney, commanding 6th Division, enjoyed success against Sir Charles Fergusson's 5th Division thanks to another text-book display of the new cavalry tactics. Colonel Hogg of the 4th Hussars had ordered his men to dismount and attack the heights of Knockanora, held by General Capper's brigade, on foot, while horse artillery poured enfilade fire onto the defenders. *The Times* noted that Hogg's troops 'thus proved most valuable to General Pulteney and showed what cavalry can do when closely co-operating with its sister arms'.[50] Hogg would command the 4th Hussars in 1914, being severely wounded and captured on 1 September at Taillefontaine, during the rearguard action at Villers-Cottérêts. He died the following day.

All of these engagements were small and strategically insignificant, but taken together, they conclusively established the tactical superiority of the British cavalry over their German counterparts. This came as no surprise to Vaughan, who later recounted that 'I knew from personal observation of them on manoeuvres that their cavalry were not as good as ours; their marksmanship was not up to our very high standard and their men were devoid of the individuality and instinct to look after himself without panic in a tight corner which distinguishes our men.'[51] One German

47 C. Byrne, *The Harp and Crown, the History of the 5th (Royal Irish) Lancers*, pp.70-73.
48 D. Kenyon, *Horsemen in No Man's Land: British Cavalry and Trench Warfare*, P.23.
49 TNA WO 279/39, Report on Army Manoeuvres 1910, p.34.
50 *The Times*, 16 September 1913.
51 Vaughan, op. cit., p.157.

Cavalry in action at the 1909 manoeuvres – a company of 7th Hussars fording the River Avon. (*The Tatler: Sporting and Country House Supplement*, 15 September 1909, author's collection)

cavalry officer was clear in his view that, thanks to the lessons taught by the Boer War and the advantages conveyed by long term service, 'the British Cavalry were indisputably far better trained for dismounted action than their Continental fellow horsemen.'[52] In this development the pre-war manoeuvres, especially those of 1912, played a significant part.

The cavalry's role in combat has attracted most attention, but it should not be forgotten that reconnaissance was an equally important part of its duties. Since the Boer War, the work of Robert Baden-Powell (Inspector-General of Cavalry) and Michael Rimington (Commander of 3rd Cavalry Brigade), which included the use of specially trained scouts, innovative training methods and a far greater emphasis on care for the horse, had resulted in a marked improvement in the performance of the cavalry in its reconnaissance role, as recognised by Sir John French during the 1906 manoeuvres and commended in the manoeuvre reports for 1912 and 1913.[53] The BEF's cavalry reconnaissance work proved itself to be clearly superior to that of the Germans in the opening weeks of the war, and Spencer Jones has credited it with keeping the army relatively unscathed during the retreat from Mons.[54] Small

52 S. Jones, 'Scouting for Soldiers: Reconnaissance and the British Cavalry, 1899-1914', *War in History,* p.512.
53 Ibid., pp.504-510.
54 Ibid., p.513.

units, mostly patrols and squadrons, were able to screen the BEF's main body through skilful reconnaissance of the advancing Germans, and this was possible thanks to the years spent training and learning from formative experiences such as Allenby's travails during the 1912 manoeuvres.

It might be thought that the Mounted Infantry's performance in September 1912 would have been enough to have secured their survival, and it has certainly been suggested that the MI could have played a valuable role for the BEF in 1914.[55] However, the success of Grierson's mixed force of cyclists, Yeomanry and Mounted Infantry in 1912 was not sufficient to prevent the MI's demise, as they were abolished the following year. There had been a recognition that the widespread assumption that the Blue mounted troops would be easily dealt with by the Red forces had been proved groundless; Grierson admitted to concerns that his mixed force would be inferior to the Red Cavalry, 'but I made the discovery on the first day that, in that particular part of the country, they were quite a match for their opponents.'[56] The Blue Cavalry Umpire Chetwode praised their fire support, powers of reconnaissance and their use for seizing tactical points, comparing them favourably with cyclists in the same role. The 1913 Field Service Manual envisaged a dual role for the MI as divisional mounted troops and as part of mixed mounted brigades. However, within a year of the East Anglia manoeuvres the Mounted Infantry had disappeared.

Andrew Winrow has argued that the Mounted Infantry could have had an important role to play when war came, arguing that 'the outcome of the 1912 Manoeuvres, at least from the perspective of the Regular Mounted Infantry, was disregarded and a potential option for the British Expeditionary Force in 1914 lost.'[57] Why was this apparent lesson of the 1912 Manoeuvres not learnt? The truth is that the experience of manoeuvres neither consigned the Mounted Infantry to extinction nor provided a convincing argument for their retention. Other considerations decided their fate.

The MI had appeared to serve a useful purpose at a time when the Regular Cavalry was equipped with inferior carbines, but by 1913 the cavalry, now equipped with the Short Magazine Lee Enfield rifle, had drastically improved its marksmanship and dismounted training. The addition of two Regular cavalry regiments from South Africa in 1912 further undermined the case for MI since the cavalry now had greater resources for supplying divisional mounted squadrons. The forces arrayed against the Mounted Infantry, whether hostility from the press or powerful voices within the Army, were simply too strong to resist, especially in light of the inherent lack of unit ethos or loyalty given that the Mounted Infantry consisted of troops taken from other infantry battalions. Once it became clear that the cavalry was now effective in

55 A. Winrow, *The British Regular Mounted Infantry 1880-1913,* DPhil thesis, University of Buckingham, 2014.
56 TNA WO 279/47 Report on the Army manoeuvres 1912, p.174.
57 Andrew Winrow in the Journal of the Society for Army Historical Research, Autumn 2015, p. 275, Journal Intelligence 1951. See also Nick Evans' rejoinder in the JSAHR 94, Spring 2016, p.72.

its developing dismounted role, the need for Mounted Infantry seemed less obvious. There were also suspicions that, however well-suited MI were to colonial campaigns and peace-time manoeuvres, they might fare less well in a continental war up against Prussian Uhlans.

However, if much was learnt from pre-war manoeuvres, many lessons were not. Many of the difficulties experienced by the cavalry in pre-war manoeuvres had been attributable to the lack of a permanent divisional staff – in 1912 Grierson had complained that no clerks or stationery were provided for the Blue Cavalry Division and had to be borrowed from Blue HQ. One of the specific recommendations made in the Manoeuvre Report – that the Cavalry Division should have a permanent staff – was not acted on, with harmful consequences two years later. Nikolas Gardner has argued that the Cavalry Division disintegrated under the pressure of the retreat from Mons thanks to the inexperience of its staff, its unwieldy size (four brigades, each of three regiments where the French and German cavalry divisions contained three, each of two regiments) and the unhelpful conduct of its brigade commanders, who took cavalry traditions of independence and initiative to dangerous levels. Edmonds was in no doubt where the blame lay: 'The truth was that Allenby's staff were useless. They never knew where his brigades were.'[58] This may be an overly harsh judgement, especially on Allenby's GSO1, John Vaughan, though it is interesting to note Hubert Gough's judgement that Allenby 'was easy to serve as long as he had a good staff officer'.[59] Despite statements to the contrary from Nikolas Gardner and the present author, Vaughan did have staff training and experience and was a product of the Staff College.[60] Vaughan himself complained that on the night before Mons 'I could get no orders for Allenby out of Archie Murray.'[61]

Even so, Allenby's premature withdrawal from the line on 24 August 1914, exposing 5th Division's left, was uncomfortably reminiscent of Vaughan's messages to Lawson's 1st Division on the morning of the 18 September 1912: 'Our Cavalry Division is now moving north-east… thus uncovering your left flank' and later, 'Can no longer protect your left.' At Le Cateau, the Division went missing just as the Red cavalry had on the final day of the 1912 manoeuvres; Edmonds visited Smith-Dorrien's Chief of Staff Brigadier-General Forestier-Walker to ask where the cavalry was, only to be told, 'I wish to God we knew.' One reason for this was that although the Cavalry Division was equipped with a wireless link to GHQ, it does not appear from the Divisional War Diary to have been used, just as had been the case in 1912. Haig had been made aware of the communications failures of the Cavalry Division shortly after the 1912 manoeuvres ended: 'Ride out and meet Colonel Greely 19th Hussars and have a talk

58 B. Gardner, *Allenby*, p.75.
59 General Sir Hubert Gough, *Soldiering On,* p.96.
60 I am grateful to Dr. Nick Evans for correcting my initial judgement of Vaughan in the *Journal of the Society for Army Historical Research* 93 (2015), pp.44-45. See his comments in the JSAHR 93 (2015) pp.273-274.
61 Vaughan, op. cit., p.159.

about the manoeuvres. He tells me that information had not been passed on from Cavalry Division to O.C. Regiments!'[62] If one of the purposes of manoeuvres was to afford commanders and staff officers realistic combat practice, the example of the cavalry staff in 1914 suggests that this was not achieved in 1912.[63]

This was part of a wider failure to put in place an effective General Staff for the BEF, and it is no coincidence that the most efficient staff work in 1914 was done by I Corps, which was the only unit to go to war with a settled staff already in place. Grierson had professed himself to be impressed with the younger generation of staff officers in 1912: 'They are extremely well trained, they are ready to take responsibility, and they are clever at interpreting the intentions of the Commander-in-Chief.'[64] Nevertheless, when war came two years later, the BEF lagged well behind its German foes and French allies in terms of having an effective staff system. The two Corps had just 31 staff officers between them (44 when one includes the artillery staff).[65] As Colonel Nicholson observed: 'Our organisation and administration on the outbreak of war may therefore be said to have started from scratch, whereas Continental Powers with their large peace armies and extensive manoeuvres had a long start.'[66]

Another area in which the British Army's manoeuvres before 1914 failed to produce much-needed improvement was in digging trenches; the Inspector-General had repeatedly complained of the standard of digging by British troops and this was still the case in August 1914, when Snow rode round 4th Division at Le Cateau to inspect the troops who were supposed to be entrenched: 'In spite of what troops had learned in the Boer War, in spite of what they had been taught, all they had done was to make a few scratchings of the nature of what was called, fifty years ago, a shelter pit, of no use whatever against any sort of fire.'[67]

Snow believed that the restrictions placed on entrenchments in pre-war manoeuvres had impeded the meaningful study of the use of ground in peace time 'and had even taught us false lessons'. However, it is interesting to learn the view from the other side – Leutnant Schacht of the Machine Gun Company of the 26th Infantry Regiment was impressed by the BEF's trench-building: 'Completely shielded by a deep gully we approached Le Cateau. We could see no trace of any trenches. The British had a brilliant understanding of the art of digging in.' Schacht observed that 'the eerie emptiness of the battlefield, which we had often discussed on quiet winter

62 Lt. Gen. D. Haig (ed. D. Scott), *The Preparatory Prologue: Diaries and Letters 1861-1914*, p.320.
63 N. Gardner, 'Command and Control in the "Great Retreat" of 1914: The Disintegration of the British Cavalry Division', *The Journal of Military History*, pp.29–54.
64 TNA WO 279/47 Report on the Army manoeuvres 1912, p.172.
65 P. Harris, *The Men who Planned the War: A Study of the Staff of the British Army on the Western Front, 1914-1918*, p.42.
66 Ibid.' p.43.
67 Lt. Gen. T. Snow (ed. D. Snow and M. Pottle), *The Confusion of Command: The War Memoirs of Lieutenant General Sir Thomas D'Oyly Snow, 1914-1915*, p.17.

Men of Grierson's Blue Army entrenched, Horseheath, September 1912. (Saanich Archives)

evenings in the mess, was very evident here. All we could make out, because of the muzzle flashes, were the enemy artillery positions.'[68]

It is routinely stated that the German army was well ahead of the British in terms of being prepared for trench warfare in 1914 – Edmonds later 'related privately' that when attending the 1908 German Manoeuvres he had witnessed 'hand-grenades, light-ball pistols, camouflage suits for wire-cutting at night'.[69] However, in reality the Germans' pre-war training had left them little better prepared than the BEF for the conditions of warfare they would encounter from the Aisne onwards. The BEF went to war in August fully expecting to have to dig – every infantry battalion took with it an Entrenching Implement (Pattern 1908) for each man in addition to 226 Shovels, 151 picks, 17 felling axes, 8 hand axes, 20 reaping hooks, 32 folding saws, 24 wire cutters, and 8 crowbars. However, as the 1911 Manual of Field Engineering reminded its readers, field fortifications were to be undertaken for the purpose of defending a particular position and regarded as a means to an end, not an end in itself, as 'field fortification presupposes a defensive attitude'. At the start of the move to trench warfare troops could have been forgiven for thinking that they were indulging in the usual process of digging temporary scrapes as they had before the war, but after October 1914 it was clear that mobile warfare was over, at least for the time being.[70]

68 N. Cave & J. Sheldon, *Le Cateau,* p.52.
69 J. Terraine, *Douglas Haig, The Educated Soldier*, p.94.
70 S. Bull, *Trench Warfare: A History of Trench Warfare on the Western Front* pp.25-26.

One practice which pre-war training, including manoeuvres, had encouraged was the preference for digging trenches on forward slopes, in order to provide a long field of fire, increasing the zone swept by fire and making the most of the defenders' small arms. This practice, which owed a great deal to lessons apparently learnt in the Russo-Japanese War, went against the *Infantry Training Manual*, which favoured placing troops in concealed trenches on reverse slopes in order to protect them from enemy artillery fire. The consequence of such faulty siting of trenches was to be seen at Ypres in October 1914, when Rawlinson, commanding IV Corps, was highly critical of Thompson Capper (7th Infantry Division) and his three Brigade commanders for the poor disposition of their troops, which resulted in heavy casualties from German shelling. Haig lamented that 'fine troops like the 7th Division' had been 'reduced to inefficiency through ignorance of their leaders in having placed them in trenches on the *forward* slopes where Enemy could see and so effectively shell them.'[71]

Tellingly, Rawlinson did not believe that it was his role to order Capper to change the dispositions, seeing it as the responsibility of the three Brigade commanders (Ruggles-Brise, Watts and Lawford). Rawlinson's GSO1, Brigadier-General R.A.K. Montgomery, was critical of the way in which pre-war doctrine had prepared 7th Division: mistakes had been made 'owing to their not being told beforehand and having followed what they had learned in their peace training, that is to say insisting on getting a good field of fire from the trenches and making everything subsidiary to that'. This was one more example of the way in which pre-war training was unable to prepare commanders for the new conditions of warfare they encountered in 1914, and the steep learning curve on which Haig, Rawlinson and the rest found themselves from the outset of the war.[72]

Manoeuvres sounded some warning bells in terms of the likely performance of troops on the march once war came. The Special Reserve had been created as part of the Haldane reforms of 1907 and provided a source of reinforcement drafts for the regular battalions. Whereas the Army Reserve contained men who had served in the army for seven years, the Special Reserve was made up of men who wanted to try army life without joining up. They enlisted for six years with an option to extend the period by another four, and after six months of training they went back to their civilian jobs but had to attend training for four weeks each year and could be called up if there was a general mobilisation.[73] The 1910 manoeuvres made heavy use of Special Reservists (there were 567 in the 1st Buffs, 569 in the 1st Gordon Highlanders, among a total of 84 officers and 4,864 NCOs and men from the Special Reserve who had been called out to bring units up to war strength), and their performance gave rise to concern on the part of Brigadier-General Ivor Maxse. He was known to be a punctilious and

71 D. Haig, *War Diaries and Letters 1914-1918*, p.75.
72 D. Filsell, 'Thompson Capper and the Forward Slopes', *The Douglas Haig Fellowship Records*, pp.29-33.
73 A. Rawson, *The British Army 1914-1918*, p.18-19.

demanding officer – Captain Brocklehurst of the 3rd Coldstream Guards complained in his diary during the 1912 manoeuvres that 'I was again on outpost all night. Maxse was very wearisome.'[74] But he was also recognised as being the army's deepest thinker on training and tactics, and he submitted a detailed report on the deficiencies of Special Reservists in the 1910 manoeuvres.[75]

Maxse described the men as 'weedy and small' and observed that their ability to cope with long marches was limited. Out of a total of 83 men of the 1st Leicestershire Regiment who fell out from the line of march on the first three days, when they marched a total of 54 miles, 79 were reservists and only four were regulars. Maxse recognised that these men were dropping out 'not from want of pluck' but due to sore feet and blisters. Some were put into boots part worn by others, some had holes in their socks and 'I personally saw many men marching with practically brand new boots.' He recommended that more care was needed in fitting boots and softening the leather. Maxse judged that 'It appears conclusively proved that reservists called up on mobilisation will, failing time for some hardening process, only at first be able to undertake very short marches.'[76] These fears were to be borne out in August 1914, when large numbers of troops, 'all of them reservists, who had not carried a pack for years', struggled to cope with marching on cobbled roads in new boots from their railheads to Mons in the August heat.[77] Maxse's complaints about the reservists' marching at a meeting with his battalion commanders on 17 August 1914 show that his forecast had been correct. This is another example of a problem being diagnosed as a result of manoeuvres but not acted on by 1914, though it is hard to see what could have been done to address the issue unless the question of training for reservists was wholly reassessed.

Another lesson not fully learnt from pre-war manoeuvres was the need for indirect artillery fire. This had been specifically commented on in Roberts' manoeuvre report as early as 1903, and Smith-Dorrien as Chief Umpire for the Blue Army in 1912 was critical of the artillery's failure to take cover. The extent of confusion over the proper use of field artillery is illustrated by the fact that only four years earlier he had urged in his report on the 1908 Aldershot Command Manoeuvres that: 'We must take more risks with our guns, pushing sections forward with the advanced infantry.'[78] Further evidence of this confusion can be seen in the fact that by 1914, the term 'close support' had been removed from British regulations and yet *Field Artillery Training 1914* stated that 'to support infantry and to enable it to effect its purpose the artillery must willingly sacrifice itself.'[79]

74 Diary of Capt. J. Brocklehurst, Coldstream Guards Archive.
75 Brig. Gen. I. Maxse papers IWM 69/53/1 File 2/1/2
76 Ibid.
77 Lord Gleichen, quoted in A. Gilbert, *Challenge of Battle*, p.47.
78 Bowman and Connelly, op. cit., p.144.
79 S. Jones, *From Boer War to World War: Tactical Reform of the British Army, 1902–1914*, p. 153.

British artillery doctrine in 1914 was, characteristically, caught between two stools and dependent on individual practice, with some favouring French tactics, emphasising rapid direct fire and close support of the infantry, while others prioritised indirect fire, slow precise ranging and concealment as their experiences in South Africa had taught them. The consequences of this lack of uniformity were to be seen at Le Cateau with Smith-Dorrien in command, when the commander of 5th Division's artillery, Brigadier-General John Headlam, deployed his guns in forward positions, fighting alongside the infantry in the manner approved by *Field Artillery Training 1914.* As a result the Division's artillery suffered heavy losses at the hands of German guns of greater firepower and range firing from behind a ridge. The Division lost 27 guns, more than a third of its total, and at least 22 officers and 180 other ranks were lost, along with 257 horses. By contrast, the artillery of 3rd and 4th Divisions, which were deployed further back and used the terrain for concealment, fared far better. The four guns that Brigadier-General Frederick Wing of 3rd Division did push forward were effectively written off as direct support guns. The power of indirect artillery fire had been conclusively proved, as had the inadvisability of the artillery being deployed alongside the infantry in the Napoleonic manner.[80] Of course it would have been preferable for the British Army for these lessons to have been learnt from manoeuvres rather than on the battlefield of Le Cateau, but such was the state of communications in 1914 that it is not surprising that the British Army struggled with the use of indirect fire at this stage in the war.

The 1912 and 1913 manoeuvres occurred at a time when the British Army was caught in an unresolved tactical debate about the impact of increased firepower and was lacking a uniform doctrine; too often the response in all three arms was to tolerate and even encourage individual commanders to use those methods which suited them. This was specifically commented on by Douglas, who observed that all four divisions in 1912 employed very different methods of attack, with 2nd Division preferring to manoeuvre an enemy out of his position by threatening his flank and line of retreat rather than deliver a decisive blow, 3rd Division deploying on a broad front and 4th Division keeping a strong reserve in hand to use once a weak spot had been located.[81] Douglas deplored this failure to establish a uniform doctrine and practice as required by *Field Service Regulations,* though elsewhere it was clearly stated that 'methods to be employed must be left to the commander on the spot.'[82] A year later Douglas was still discerning a lack of uniform doctrine among commanders while admitting that it was unrealistic to expect much to have changed in the space of twelve months.[83] Of course, this tendency was not unique to the British Army – after inspecting II Army Corps in Pomerania in the autumn of 1905, Kaiser Wilhelm contrasted the tactics

80 S. Marble, *British Artillery on the Western Front in the First World War,* p.44 48.
81 TNA WO 27/508, Report of the Inspector-General 1912 p.8.
82 Bowman and M. Connelly, *The Edwardian Army,* p.74.
83 NA WO 27/510 Annual Report of the Inspector-General 1913, p.6.

An 18 pounder gun in action in a forward position, September 1912. (Saanich Archives)

employed by its commander General Langenbeck with those encountered elsewhere, observing that: 'When I'm in East Prussia I find one tactic, when in Metz another – and when I go to Hanover I find something else entirely – which is itself different from Silesia.'[84]

However, there was perhaps less excuse for such lack of uniformity in an army as small as Britain's. There has been a great deal of debate over the extent to which the British did, in fact, have an identifiable doctrine before 1914, with Stephen Badsey arguing persuasively that even if a formal doctrine was lacking, that does not mean that one did not exist in the form of the many manoeuvre reports, drill books and training manuals produced by the army.[85] It can also be argued that in many ways this flexible and non-formal approach would stand the British Army in good stead during the First World War, for example in its development of the tank, which was far in advance of anything the German army, far more formal in its learning processes, was capable of.[86]

Martin Samuels has argued that the British Army suffered from a command system known as 'umpiring' whereby commanders gave general orders and then abdicated

84 E. Dorn Brose, *The Kaiser's Army,* p.153.
85 S. Badsey, *Doctrine and Reform in the British Cavalry 1880-1918*, p.3.
86 R. Foley, Dumb donkeys or cunning foxes? Learning in the British and German armies during the Great War, *International Affairs*, pp.291-293.

responsibility for looking after the details to their subordinates, limiting themselves to general guidance. Haig's Chief of Staff Johnnie Gough complained that: 'Some commanders are so anxious to leave subordinates a free hand that they forget their own duty of control and guidance while others go to the opposite extreme and interfere with the methods of their subordinates. It does not appear easy to hit off the happy mean.'[87] We have already seen something of this in the episode of the siting of trenches on forward slopes by the three brigades in 7th Division in October 1914. The effect of this approach, it is argued, was to turn senior commanders into observers rather than participants, something of which the likes of French and Smith-Dorrien already had extensive experience as directors and umpires (in the original sense of the word) at manoeuvres.[88] Martin Samuels goes as far as to suggest that a British commander operating under this command system would abdicate his command responsibilities and not interfere with subordinates. As Haig stated: 'The chief duty of the higher command is to prepare for battle, not to execute on the battlefield. After having clearly indicated to subordinate leaders their respective missions, we must leave the execution to them.'[89]

This tendency would later have serious consequences on the Western Front, when a culture of decentralisation and individual initiative meant that there was a reluctance to share experience of methods which worked well in the trenches. Rawlinson's orders for the offensive on 1 July 1916, the first day of the Battle of the Somme, were so ambiguous that they granted his corps commanders considerable discretion, and some in turn chose to delegate the decisions to those below them. To take one example, the troops did not all go at a slow pace across no man's land on 1 July as is so often claimed. While VIII Corps' commander, Hunter-Weston, warned his troops not to advance at the double except for short distances, some battalions from each of his two divisions did attempt to rush German positions, and so did Morland's X Corps and Horne's XV Corps, while both divisions of Pulteney's III Corps seem to have advanced at a slow walk (and suffered the greatest number of deaths on 1 July).[90] As Gary Sheffield has shown, brigades within the same division could easily adopt different methods, each of them adapting to local factors and their commander's tactical precepts.[91] One brigade commander in Morland's Corps, Brigadier-General James Jardine (97th Brigade in 32nd Division), drew on his experience of Japanese infantry tactics in Manchuria to urge Rawlinson to move the infantry to within 30

87 Ian F W Beckett, 'Hubert Gough, Neill Malcolm and Command on the Western Front' in Brian Bond et al, *Look to your Front – Studies in the First World War by the British Commission for Military History*, p.7.

88 M. Samuels, *Command or Control?: Command, Training and Tactics in the British and German Armies, 1888-1918*, p.49.

89 Ibid. p.34.

90 R. Prior and T. Wilson, *Command on the Western Front: The Military Career of Sir Henry Rawlinson 1914-1918*, p.159.

91 G. Sheffield, *Forgotten Victory, The First World War: Myths and Realities*, p.167.

or 40 yards of the barrage rather than the 100 yards being practised; when met with disdain from Rawlinson, Jardine persisted in ordering his two lead battalions into No Man's Land at 7:23 a.m. on the day of the attack, so that their first waves had only 40 yards to cross when the barrage lifted.[92]

A similar degree of discretion was allowed to corps and divisional commanders in deciding the artillery bombardment to be employed, with Morland and Pulteney and the divisional commanders of VIII Corps all ignoring instructions and firing barrages which jumped rapidly from one trench to another, whereas in XIII Corps (Congreve) and XV Corps a creeping barrage was employed, albeit with insufficient field guns per yards of trench attacked.[93]

The 'carnage' at Horseheath in 1912 and the similarly bloody combat at Sharman's Hill in 1913 demonstrated that the British Army was still struggling with the central debate over whether infantry should employ dense firing lines, inviting enormous casualties, or extend the line and risk losing firepower. One of the most frequent complaints about British manoeuvres of the period was that the infantry tended to ignore the devastating firepower of artillery and machine guns, and the same problem was evident in Austro-Hungarian manoeuvres, where umpires were under orders to downplay the effects of firepower in order to maintain morale and inculcate the 'offensive spirit.' As a result, infantry assaults were allowed to proceed at an unrealistic rate and to continue when they would have been halted by prohibitive casualties in real life.[94] At the Aisne in September 1914 the BEF sustained crippling casualties in a series of frontal assaults on strong defensive positions, although from the start of the war it was clear that the British commitment to attacking in dense formations was less extreme than the Austrians, the Germans or the French; this was due as much to their experience in South Africa as to their pre-war manoeuvres.

Finally, manoeuvres raised significant question marks over those who would lead the army in the next war. Plumer was heavily criticised for his leadership of the Red army in 1910 – 'in the execution of his plan of envelopment the Red Commander took great and unnecessary risks' – but this did not prevent him from eventually rising to command an army on the Western Front, even if when war broke out he was somewhat becalmed at Northern Command, only arriving in France in February 1915. The 1912 Manoeuvres featured disappointing performances from Haig and, in particular, Allenby, and French made a poor impression the following year. If this was due to lack of experience in handling large bodies of troops, as Douglas argued, then surely this demonstrated the value of such manoeuvres in affording them such experience. He had defended their performance in the following terms:

92 Samuels, op. cit., p.144.
93 Prior and Wilson, op. cit., p.165.
94 L. Sondhaus, *Franz Conrad von Hötzendorf: Architect of the Apocalypse*, p.92.

A machine gun of the Blue Army in action, September 1912. (Saanich Archives)

> Their failures, such as they were, were due to lack of experience in the actual handling of large bodies of troops. This lack of experience in the handling of forces of all arms is general throughout the higher ranks of the Army. The value of the system of manoeuvres now held annually is thus clearly demonstrated.[95]

However, the fact that Haig and Allenby both made some of the same mistakes again in August 1914, and for some of the same reasons, calls into question how much either had learnt from the experience of 1912.

The conditions of warfare in 1914, both in its frenetic mobile phase and then following the development of trench warfare, imposed physical and psychological strains on commanders for which peacetime manoeuvres, however realistic, could scarcely prepare them. Grierson's death is the most dramatic example, but a significant number of senior staff officers collapsed under the strain of the retreat from Mons, notably Murray and Vaughan (both on 26 August) and Edmonds (30 August); the latter was replaced as 4th Division's Chief of Staff on 4 September and was subsequently found a less taxing staff job at GHQ. Murray's collapse was described in the following terms by Edward Spears:

95 TNA WO 27/508, Annual Report of the Inspector-General 1912, pp.4-5.

> The strain had been so frightful that the Chief of Staff, Sir Archibald Murray, exhausted by anxiety and overwork, had had a temporary collapse from shock when false news arrived that the enemy had attacked and defeated the I Corps at Landrecies. Colonel Cummins, the MO at GHQ, was very anxious about him. He did not actually faint, but his pulse was very weak.[96]

Murray – described by Haig as 'a kindly fellow, but not a practical man in the field' – was revived with an injection from Cummins ('morphia or some other drug', according to Wilson). Murray himself recorded that the 'Doctor injected something that pulled me round for the day',[97] but his incapacity at such a critical moment caused chaos in the army's staff work, and such was the lack of confidence in him that it was a relief when he was sent on sick leave in January 1915 and then replaced by Robertson.[98] An even more serious collapse than Murray's was that of Colonel Frank Boileau, Chief of Staff of 3rd Division, who broke down under the strain and shot himself on 27 August, dying the following day; Captain H.B. Owens, Medical Officer to the 3rd Cavalry Brigade, reported that he had 'had to see a staff officer who had just shot himself in the head with a revolver in a motor car. He was still living.' Boileau's suicide was hushed up, for understandable reasons at such a critical point in the retreat.[99] His successor Frederick Maurice, who wrote home that 'poor old Boileau went quite off his head with the strain', recognised the enormous pressure on the staff: 'The work of organising retirements is about the most difficult job a staff can have to do.'[100]

In such testing conditions, with chronic lack of sleep adding to the effects of a demoralising retreat, serious mistakes could be made (French later wrote of this period that 'some of the Headquarters Staff had not closed their eyes for 48 hours').[101] 2nd Division was commanded by Lieutenant-General Charles Monro, whose Brown force had staged the fighting retreat in the 1913 manoeuvres a year before. Following the fierce rearguard action at Villers-Cotterêts on 1 September, Monro gave the order to Major George Jeffreys of the 2nd Grenadier Guards to open fire on some cavalry which had appeared out of the forest 500-600 yards away. Having inspected the approaching horsemen with his field glasses, Jeffreys saw from the colour of their horses that they were the Scots Greys (who had been part of Monro's force the year before) and persuaded Monro to change the order. As Jeffreys observed in his diary, 'I think he was tired and overwrought.'[102]

96 E.L. Spears, *Liaison 1914*, p.239.
97 J. Bourne, 'Major General Sir Archibald Murray' in S. Jones (ed.), *Stemming the Tide*, p.64.
98 Holmes, *The Little Field Marshal*, pp.266-268, K. Jeffrey, *Field Marshal Sir Henry Wilson: A Political Soldier* pp.134-137.
99 Gilbert, op. cit., p.155.
100 Harris, op. cit., pp.44,45.
101 Field-Marshal Viscount French of Ypres, *1914*, p.69.
102 L. Macdonald, *1914*, p.268.

Even the proverbially robust Sir Douglas Haig had a momentary 'wobble' at Landrecies on 26 August, though Gary Sheffield has argued convincingly that this did not amount to the 'panic' suggested.[103] After the transition to trench warfare several generals who had made a good impression in 1912, such as Snow and Morland, struggled under new and unforeseen conditions, which suggests that peacetime manoeuvres had not been ideal preparation (the same could, of course, be said for the French and German armies). It should be noted, however, that Morland was respected by his officers and highly rated as a corps commander by Haig, even if his performance in battle was noticeably uneven; it is much harder to find anyone with anything good to say about Snow's performance on the Western Front.[104]

Two other generals deemed to have performed well in pre-war exercises were Brigadier-General Ivor Maxse and the New Zealander Brigadier-General R.H. Davies, both of whom were prominently involved in the 1912 and 1913 manoeuvres and served in Haig's Aldershot command. Yet both were deemed disappointments in the opening weeks of the war, primarily because of a perceived lack of offensive spirit. Haig recorded in his diary of 6 September that Maxse, commanding the 1st (Guards) Brigade, 'seemed to have lost his fighting spirit which used to be so noticeable at Aldershot in peace time!' As a result one of the foremost tacticians in the army was sent home to command the newly-formed 18th Division of the New Army. Two days earlier Haig had told his wife in a letter that 'little New Zealand Davies has disappointed me; the want of sleep and food has quite changed him.'[105] Davies was removed from command of 6th Brigade on 22 September and seems never to have recovered from the strain of the retreat, which was exacerbated by his insistence on always marching at the head of his brigade. He was sent home and given command of the newly raised 20th (Light) Division, but led it in only one minor engagement, at Fromelles in September 1915, before handing over command in the spring of 1916 due to health issues. After a period as GOC of a reserve centre at Cannock Chase in Staffordshire, he was relieved of command for a third time and committed suicide in May 1918 at the Special Neurological Hospital for Officers in Kensington.

Gardner has argued that those officers whose careers prospered in the new conditions from the Battle of the Aisne onwards were the ones placing a heavy emphasis on courage, physical exercise and the maintenance of morale among the rank and file.[106] This was often manifested in a fondness for costly trench raids and an insistence on holding onto ground for morale purposes. These men, like Haig, were 'hybrid'

103 G. Sheffield, *Douglas Haig: From the Somme to Victory,* pp.80-83. See also E. Greenhalgh, 'Did Sir Douglas Haig panic at Landrecies in August 1914 during the Great Retreat?', *Journal of the Society for Army Historical Research,* pp.226-240.
104 W. Thompson, *Morland – Great War Corps Commander: War Diaries & Letters, 1914-1918,* p.xxvii.
105 N. Gardner, *Trial by Fire: Command and the British Expeditionary Force in 1914,* p.97.
106 N. Gardner, *The Beginning of the Learning Curve: British Officers and the Advent of Trench Warfare, September-October 1914,* pp.14-16, 19.

officers, ready to embrace technological change and tactical developments but also convinced of the importance of physical courage and personal leadership on the battlefield. Perhaps these were not qualities for which the pre-war manoeuvres had given much scope.

However, it is certainly arguable that the work that the British Army had done in 1913 and earlier in 1914 on practising a retreat under fire and the fact that a number of senior officers, such as Robertson, Haig and Snow, had studied the issue before the war played a considerable part in helping to keep the BEF together under the appalling strain of a harrowing 13 day retreat covering nearly 200 miles, pursued all the way by the Germans.

Many of those involved in the retreat from Mons, including Lomax (Commander of 1st Division), Haking (Commander of 5th Brigade) and Forestier-Walker (Chief of Staff of II Corps) had taken part in a significant Staff Ride organised by Haig in October 1907, when about 50 officers had been confronted with an exercise in the enclosed and hilly area around Monmouth which involved one of the two sides, 'Eastland', in a long and testing retreat. As had been noted at the time, such a retreat 'tries "moral" (sic) severely, and commanders have often been forced to fight merely to put better heart into their men', an interesting observation in the light of Smith-Dorrien's decision to stand and fight at Le Cateau. The report on the Staff Ride suggested that all officers involved in such a retreat 'should do their utmost to set a good example of patience and cheeriness under such difficulties in order to prevent any serious deterioration of "moral"'. According to Andrew Wiest, Haig did precisely that in August 1914: 'Haig remained in firm control, and the BEF succeeded in executing one of the most difficult of all military manoeuvres – a retreat under fire.'[107] The same Staff Ride had also addressed the issue of coalition warfare, as Eastland had been confronted with two enemies; the General Idea had noted beforehand that 'there will certainly be hesitation in the case of both Westland and Northland to place a portion of their respective armies under the command of the Commander-in-Chief of the other country', something which Haig would definitely experience for himself on the Western Front.[108]

Having spent so long considering the part manoeuvres and exercises played in shaping the performance of the BEF in the opening months of the war, it should not be forgotten that another factor was the crucial underpinning provided by the invaluable experience of those officers and men who had fought in Britain's many colonial campaigns, and in particular in South Africa. Thanks to the multiplicity of Britain's imperial commitments, the BEF's commanders had a wide range of combat experience in a variety of terrains and climates, even if most of these had been 'small wars' with little apparent relevance to the conditions they encountered in France and

107 A. Wiest, *Haig: The Evolution of a Commander*, p.17.
108 TNA WO 279/17, Report on a Staff Ride Held by the Chief of the General Staff, October, 1907.

Belgium. Lieutenant-Colonel F.G. Anley of the 2nd Essex Regiment had served in Sudan (1884-5), Dongola (1896), the Nile (1897, 1898 and 1899) and South Africa (1899-1902) while Lieutenant-Colonel H.R. Davies of the 2nd Ox and Bucks had served in Burma (1887-8, the North-West frontier (1897-8), Tirah (1897-8, China (1900) as well as South Africa.[109] Of the 157 men commanding Regular battalions in 1914, 61 had experience in more than one campaign and only 19 had no previous war service. The one war in which they had been confronted with an enemy equipped with modern weapons had been the Boer War. Seventy-one per cent of regular commanding officers had war experience from South Africa, most of them having been at an ideal age, when serving as lieutenants and captains, to be receptive to new ideas and tactics.

If the battalion commanders had a level of experience beyond their German counterparts, their men had a similar leavening of experience in the form of the Boer War veterans to be found among the ranks of the BEF, including the reservists who made up as much as 60% of many battalions. The 4th Royal Fusiliers had to call up 734 reservists in order to reach its war establishment of just over 1,000 officers and men.[110] The experience of these veterans would be especially valuable when it came to infantry rifle fire in 1914. The effectiveness of Boer musketry had profoundly shocked the British, leaving a powerful impression, and this led to important reforms which made the BEF legendary for its own rifle skills from the first clashes with the Germans.

After 1902, the old stress on volley firing was abandoned and soldiers were encouraged to take a pride in their own marksmanship, which was encouraged by fiercely contested rifle competitions, badges for marksmanship and extra pay for those reaching the required standard. This was true of the cavalry as much as it was of the infantry. The stress was on rapid firing as well as accuracy, and the epitome of this approach was the 'Mad Minute', in which a soldier had to fire fifteen aimed rounds at a target three hundred or more yards away within sixty seconds.[111] The fruits of all this hard work since 1902 were to be seen in the first three months of the war, when well-trained veterans firing from concealed positions inflicted crippling losses on German troops advancing in close order. According to John Lucy of the Royal Irish Rifles, the Germans at Ypres in November 1914 were 'badly directed, and their men not yet extended in lines. What tactics! We let them have it. We blasted and blew them to death.'[112]

The accuracy of the British musketry was much feared by the Germans. As one survivor of Le Cateau recalled: 'Unthinking, section after section ran into the well-directed fire of experienced troops. Every effort had been put into our training, but

109 P. Hodgkinson, 'The Infantry Battalion Commanding Officers of the B.E.F.' in S. Jones (ed.), *Stemming the Tide*, p.299.
110 Gilbert, op. cit., p.29.
111 S. Jones, 'Shooting Power': A Study of the Effectiveness of Boer and British Rifle Fire, 1899-1914, *British Journal for Military History*, p.42-43.
112 J. Lucy, *There's a Devil in the Drum*, pp.224-225.

it was completely inadequate preparation for such a serious assault on battle-hardened, long service colonial soldiers.'[113] Walter Bloem, an officer in the Brandenburg Grenadiers, was struck by the British use of cover at Mons: 'Wonderful, as we marched on, how they had converted every house and wall into a little fortress: the experience, no doubt, of old soldiers gained in a dozen colonial wars; possibly even some of the butchers of the Boers were amongst them.'[114] An officer from the 136th Infantry Regiment at Ypres was stunned by the accuracy of a British soldier who swung round in a standing position and shot at the German's comrade, wounding him in the head: 'This accuracy shown by the long service British soldiers with colonial experience who were deployed opposite the company, verged on the miraculous.'[115] This became something of an *idee fixe* among the Germans, who were apt to assume that they were confronted by Boer War veterans when there was usually no way of knowing how old or experienced the British troops opposite them were.

113 Jones, op. cit., p.45.
114 A. Mallinson, *1914: Fight the Good Fight: Britain, the Army and the Coming of the First World War,* p.60.
115 Jones, op. cit., pp.45-46.

# 8

# Conclusion

When Colonel Huguet took up his position as French Military Attaché in London in 1905, his predecessor 'did not conceal from me his opinion that the Army, to which he had been attached in the Boer War, was only fit for police duties or for minor colonial expeditions'. Initially convinced that this view was correct, Huguet had decided by 1914 that the British Army had been transformed; it could now be 'a substantial help when the war between France and Germany broke out, and it was certain that this would soon come to pass... Russia would be on our side. English military assistance would be very valuable also and it was in our interests to do all in our power to assure it.'[1] Having attended the 1912 manoeuvres, General Foch agreed with his countryman, reporting back to the French War Minister: 'English army: first quality; young, well-informed if small command.'[2] The French General Staff's official report on the same manoeuvres was equally positive: 'The English army constitutes a well-recruited force with all the qualities of professional soldiers, with rich cadres of well instructed and experienced NCOs, a corps of carefully trained officers, and battle hardened by the ordeals of the South African war.'[3]

It is clear that the British Army had made enormous strides between the conclusion of the Boer War in 1902 and the outbreak of European war in 1914. Contemporary judgements paid tribute to the role played in this progress by training and manoeuvres, with Haig's use of staff rides singled out for particular praise.[4] One regimental history stated:

> The time between the close of the South African War and the outbreak of the Great War was one of intense change in the training and organisation of the Army. Lessons learnt in the South African War, and those derived from the

1 General V. Huguet, *Britain and the War: A French Indictment*, pp.3-4.
2 D. Porch, *The March to the Marne: The French Army 1871-1914*, p.228.
3 Report of the General Staff, Revue militaire des armées étrangères 1913, p.34.
4 General Sir Hubert Gough, *Soldiering On*, p.95.

> study of the Russo-Japanese War, were absorbed and reduced to a theory of warfare suitable for the role likely to be filled in the future by the British Army… … Annual manoeuvres on a large scale were now introduced as a regular force of training.[5]

There seemed to be grounds for confidence for the coming conflict. Grierson, one of the central figures in pre-war British manoeuvres, faced the future with guarded optimism, writing in December 1912 to his kinsman, the Rev. Robert Grierson: 'I think we are fairly well prepared – at least better than we have ever been before.'[6] The Canadian politician Sam Hughes told his audience at the Empire Club of Canada on 14 November 1913: 'The British soldier of today, I am satisfied, is far ahead of what he was at the time of the South African War, and then I think he was equal to anything that had been in history previously.'[7] Hughes was, of course, speaking with political motives in mind, but he was certainly in a strong position to judge the transformation, as he had both fought in South Africa and witnessed the 1912 manoeuvres. Colonel Richard Meinertzhagen, a British officer of German descent, noted in his diary at the start of the war: 'Our Expeditionary Force is terribly small, but a mighty weapon, for every soldier can shoot and every man is determined to fight. The Germans will soon find that out. We are not the soldiers of the South African War.'[8]

In this process of reform, manoeuvres played their part along with exercises, war games, staff rides and all the other forms of training the army underwent each year, but so too did combat experience from imperial conflicts, above all South Africa. There was a limit to what could be learnt from pre-war manoeuvres, but these limits were less the product of conservatism or lack of imagination on the part of the generals and more the result of the restrictions imposed on the manoeuvres in terms of money, numbers and available space. In particular, the budgetary limitations set by a government for whom the Royal Navy was always the priority frequently interfered with operational considerations, as in 1903 when the cancellation of the Army manoeuvres was considered in order to cut costs and in 1904 when the commanders' wish to practise an opposed landing was ignored. The Army's leaders had to operate in a political culture where there was widespread acceptance of Admiral Fisher's concern that 'every penny spent on the Army is a penny taken from the Navy'.[9] Whereas the Naval Estimates rose from £32.9 million in 1907-08 to £46.3 million in 1913-14, the Army Estimates only rose from £27.3 million in 1909-10 to £27.6 million in 1911-12 and £28 million in 1912-13.[10] William Robertson went so far as to claim that the

5 Major A. Whitehorne, *The History of the Welch Regiment*, p.263.
6 D. Macdiarmid, *The Life of Lieut. General Sir James Moncrieff Grierson*, p.251.
7 A. Capon, *His Faults Lie Gently – The Incredible Sam Hughes*, p.87.
8 S. Jones, 'Shooting Power: A Study of the Effectiveness of Boer and British Rifle Fire, 1899-1914', *British Journal for Military History*, p.43.
9 N.A. Lambert, *Sir John Fisher's Naval Revolution*, p.88.
10 Col. J. Dunlop, *The Development of the British Army 1899-1914*, p.301.

priority given to the Royal Navy meant that the Army was denied equipment which it knew to be necessary.[11] In addition the pre-war Army was serving governments of both parties which refused to face harsh strategic realities, above all the fact that in the absence of conscription the British could never hope to compete with the continental armies in terms of numbers.

Contrary to widespread assumptions, the British Army's pre-war manoeuvres, and in particular those of 1912 and 1913, did teach the Army useful lessons, such as the value (and limitations) of aerial reconnaissance and the proper use of cyclists and cavalry. Within the many limitations inherent in such an exercise, 1912 in particular constituted a fair approximation of the type of fighting experienced in the opening weeks of the Great War. There was actually a strong resemblance between the events of 16-18 September 1912 and the encounter battles of Mons and Le Cateau, even if Mons' terrain was very different. The fact that these two battles, fought just three days and 30 miles apart, were fought in such contrasting terrain – Mons an urban, industrial landscape of a sort rarely found outside the Midlands or Northern England, Le Cateau open, rolling farmland more like South Cambridgeshire – shows just how difficult it could be to make any peace-time exercise a realistic preparation for war conditions. There was the further difficulty that after a decade of practising offensives, all the fighting until the counter-attack at the Marne only involved the BEF in defending and retreating.

The tactical problems covered in the previous decade had included an amphibious landing (1904), an encounter battle (1912), and an assault on entrenched positions along with a fighting retreat (1913), a programme which constituted a more varied and worthwhile range of exercises than their continental counterparts were attempting. Transport, supply and medical arrangements had been thoroughly tested and new technological developments, such as the use of aircraft, wireless telegraphy, telephones, transport lorries and even tracked vehicles, had been tried out.[12] Motor cycles had been employed on manoeuvres since 1903 and had proven their worth.[13] Valuable experience of command at corps level had been afforded to Haig, Plumer and Smith-Dorrien, at divisional level to future army commanders such as Rawlinson and Allenby, and at brigade level to the likes of Maxse, Cavan and Haking, each of whom would rise to corps command on the Western Front. Maxse, in particular, would have an important part to play in the BEF's tactical development as the war went on.

The British went to war in August 1914 with a small but well trained army whose greatest strengths were its flexible doctrinal tradition and its excellent marksmanship.

11 Field-Marshal Sir W. Robertson, *From Private to Field-Marshal*, pp.192-3.

12 For developments in communications, see N. Evans, 'The British Army and communications, 1899-1914', *Journal of the Society for Army Historical Research*, pp.208-224.

13 For the vital contribution of the BEF's despatch riders in 1914, see M. Carragher, 'Amateurs at a professional game': The Despatch Rider Corps in 1914 in S. Jones (ed.), *Stemming the Tide*, pp.332-349.

However, there were also serious deficiencies, not least the fact that the General Staff, created less than a decade before, lacked the experience and stature of its continental counterparts. The BEF's greatest weakness, other than its size, was its lack of heavy artillery, and in particular of heavy field howitzers, though its combination of field guns, field howitzers and heavy guns was tactically superior to the French, who were almost entirely reliant on the 75mm gun. It is also worth remembering that the artillery, like so many other features of the 1914 BEF, reflected the nature of the army as a mobile strike force, which is why 'galloper guns' such as the 18 pounder suited it more than heavy howitzers.

Pre-war manoeuvres showed the British Army to have made significant advances in professionalism and competence since the Boer War. The problem was not that the manoeuvres carried out by the British Army in the years before 1914 failed to prepare it for the kind of warfare it would experience when war broke out. The problem was that this phase of mobile warfare would be short-lived, and would give way after October 1914 to static trench warfare for which the British were no better or worse prepared than the armies of France or Germany. This presented a whole new set of tactical problems that would take the British Army several years to come to terms with.

It is a staple of recent writing on the British army in the Great War that after 1916 it was engaged in a steep 'learning curve' which enabled, and culminated in, the victories of the final months of the war. This is now more properly referred to as a 'learning process', in recognition of the fact that it was neither smooth nor uniform. As Gary Sheffield has observed: 'In reality there was not one learning curve, but several.'[14] Undoubtedly, the BEF did learn from its experiences earlier in the war to become an effective unit in the final, more mobile phase of the fighting on the Western Front, even if tactical development had been frustratingly slow and sporadic in the first two years of the war. Improvements in training were crucial to this process. The Somme, in particular, is held to have taught lessons which laid the foundations for the victories of the 'Hundred Days', though it should not be forgotten that simultaneously the Germans were also learning from their own experiences of trench warfare, above all the need for a more flexible defensive system. But in reality, all armies are permanently on a learning curve, generally steeper and more urgent in the intense conditions of war time, and the British Army was already on such a curve following the Boer War when it was called into action in 1914; in this process the manoeuvres and exercises of the previous decade played a crucial part. British army manoeuvres between 1903 and 1914 had certainly not been a futile exercise.

14 G. Sheffield, 'The Somme: A Terrible Learning Curve', *BBC History Magazine*, July 2011. See also W. Philpott, 'Beyond the Learning Curve: The British Army's Military Transformation in the First World War', RUSI Website, 10 November 2009. https://rusi.org/commentary/beyond-learning-curve-british-armys-military-transformation-first-world-war (accessed 30/8/15).

# Appendix

## Order of Battle for the 1912 Army Manoeuvres

---

### Red (Aldershot Command)

Commander-in-Chief: Lieutenant-General Sir Douglas Haig
Chief of Staff: Brigadier-General Francis Davies

**Cavalry Division**
GOC: Major-General Edmund Allenby

*1st Brigade:*
GOC: Brigadier-General Charles Kavanagh
2nd Dragoon Guards
19th Hussars

*2nd Brigade:*
GOC: Brigadier-General Henry de Beauvoir de Lisle
4th Dragoon Guards
16th Lancers
18th Hussars

*4th Brigade:*
GOC: Brigadier-General Charles Bingham
3rd Hussars
9th Lancers
20th Hussars

**1st Infantry Division**
GOC: Major-General Samuel Lomax

*1st Brigade:*
GOC: Brigadier-General Ivor Maxse

3rd Coldstream Guards
2nd Scots Guards
1st Norfolk Regiment
1st Cameron Highlanders

*2nd Brigade:*
GOC: Brigadier-General Thomas Morland
2nd York and Lancaster Regiment
2nd Royal Sussex Regiment
1st Dorsetshire Regiment
1st Loyal North Lancashire Regiment

*3rd Brigade:*
GOC: Brigadier-General Herman Landon
1st Somerset Light Infantry
2nd Border Regiment
2nd Essex Regiment
2nd Middlesex Regiment

**2nd Division**
GOC: Major-General Henry Lawson

*4th (Guards) Brigade:*
GOC: Brigadier-General Lord Cavan
2nd Grenadier Guards
3rd Grenadier Guards
1st Coldstream Guards
2nd Coldstream Guards

*5th Brigade:*
GOC: Brigadier-General Richard Haking
4th Royal Fusiliers
2nd Suffolk Regiment
1st Bedfordshire Regiment
2nd Oxfordshire and Buckinghamshire Light Infantry

*6th Brigade:*
GOC: Brigadier-General Richard Davies
1st Leicestershire Regiment
2nd Royal Inniskilling Fusiliers
1st Hampshire Regiment
1st King's Royal Rifle Corps

**Blue (Southern and Eastern Commands)**

Commander-in-Chief: Lieutenant-General Sir James Grierson
Chief of Staff: Brigadier-General Robert Montgomery

**Cavalry**
GOC: Colonel Charles Briggs

*Regular Mounted Brigade:*
GOC: Colonel Edward Brown-Synge-Hutchinson VC
Composite Regiment, Household Cavalry
2nd Dragoons
Mounted Infantry Battalion

*1st S.W. Mounted Brigade:*
GOC: Colonel Herman Le Roy-Lewis
Royal Wiltshire Yeomanry
Hampshire Yeomanry
Dorset Yeomanry

**3rd Division**
GOC: Major-General Sir Henry Rawlinson

*7th Brigade:*
GOC: Brigadier-General Laurence Drummond
2nd Lancashire Fusiliers
3rd Worcestershire Regiment
1st Duke of Cornwall's Light Infantry
2nd Royal Munster Fusiliers

*8th Brigade:*
GOC: Brigadier-General Beauchamp Doran
2nd Royal Scots (Lothian Regiment)
2nd Sherwood Foresters (Nottinghamshire and Derbyshire Regiment)
1st Northamptonshire Regiment
4th The Duke of Cambridge's Own (Middlesex Regiment)

*9th Brigade:*
GOC: Brigadier-General Herbert Burney
2nd Royal Warwickshire Regiment
1st Gloucestershire Regiment
1st Worcestershire Regiment
2nd Wiltshire Regiment

**4th Division**
GOC: Major-General Thomas D'Oyly Snow

*10th Brigade:*
GOC: Brigadier-General Aylmer Haldane
2nd King's Royal Rifle Corps
2nd Seaforth Highlanders
1st Royal Irish Fusiliers
2nd Royal Dublin Fusiliers

*11th Brigade:*
GOC: Brigadier-General Henry Heath
1st East Lancashire Regiment
2nd Durham Light Infantry
1st Gordon Highlanders
1st Rifle Brigade

*12th Brigade:*
GOC: Brigadier-General Henry (H.F.M.) Wilson
2nd Royal Lancashire Regiment
1st South Wales Borderers
1st Princess Charlotte of Wales's Own (Royal Berkshire Regiment)
2nd Royal Irish Rifles

**Territorial Detachment**

GOC: Major-General Walter Lindsay
*Liverpool Infantry Brigade:*
GOC: Colonel A.R. Gilbert
5th Liverpool Regiment
6th Liverpool Regiment
7th Liverpool Regiment
8th Liverpool Regiment

| | Red | Blue | Total |
|---|---|---|---|
| Men (inc. Cavalry & Cyclists) | 22,558 | 24,921 | 47,479 |
| Horses | 7,355 | 5,409 | 12,754 |
| Cavalry | 2,889 | 1,745 | 4,634 |
| 13 Pounders | 12 | 3 | 15 |
| 18 / 15 Pounders | 54 | 66 | 120 |
| 4.5 Inch Howitzers | 18 | 18 | 36 |
| 6 Inch Howitzers | 4 | 4 | 8 |
| 60 Pounders | 8 | 8 | 16 |
| Machine guns | 64 | 67 | 132 |
| Cyclists | 1,384 | 1,477 | 2,861 |
| Motorcycles | 80 | 88 | 168 |
| Aeroplanes | 6 | 9 | 15 |
| Airships | 1 | 2 | 3 |

# Bibliography

## UNPUBLISHED SOURCES


**The National Archives (TNA), Kew**
WO 108/184 Grierson's Notes on the South African War
WO 27/507 French Army Manoeuvres 1906
WO 27/508 Report of the Inspector-General of Forces 1912
WO 27/509 Report of the Inspector-General of Forces 1913
WO 275/510 Report of the Inspector-General of Forces 1908
WO 279/7 Army Manoeuvres 1903
WO 279/8 Army Manoeuvres 1904
WO 279/10 Eastern Command Manoeuvres 1906
WO 279/17 Staff Ride held by the Chief of the General Staff 1907
WO 279/18 Report of a Conference of General Staff Officers 1908
WO 279/21 Aldershot Command Staff Tour and Manoeuvres 1908
WO 279/25 Report of a Conference of General Staff Officers 1909
WO 279/31 Army Manoeuvres 1909
WO 279/39 Army Manoeuvres 1910
WO 279/40 Irish Command Manoeuvres 1910
WO 279/47 Army Manoeuvres 1912
WO 279/48 Report of a Conference of General Staff Officers 1913
WO 279/52 Army Exercise 1913
WO 33/364 Record of a Strategic War Game 1905
CAB 45/129 Account of the Retreat of 1914 by General T D'O Snow

**The National Library of Scotland, Edinburgh**
Gen. Sir Aylmer Haldane, *Diary 1875 to 1913,* MS. 20247 (i-415)

**Coldstream Guards Archive, Wellington Barracks**
Captain C.M. Pereira diary
Captain J.H. Brocklehurst diary

**Liddell Hart Centre for Military Archives, King's College, London**
Brigadier-General Sir J.E. Edmonds, *Memoirs*, Edmonds Papers

**Nuffield College Library**
The papers of John Edward Bernard Seely, Lord Mottistone

**Newspapers and Journals**
*The Cambridge Daily News*
*The Daily Mail*
*The Daily Telegraph*
*East Essex Advertiser and Clacton News*
*The Essex Newsman*
*The Evening News*
*The Graphic*
*The Manchester Guardian*
*The Morning Post*
*The Nelson Colonist*
*The New York Herald*
*The New York Times*
*The Northampton Mercury*
*The Pall Mall Gazette*
*The South West Suffolk Echo*
*The Times*
*The Westminster Gazette*
*The Wolverton Express*

## THESES

Christopher Yi-Han Choy, *British War Gaming, 1870-1914*, MA Thesis in Conflict Studies, King's College London, 2013.

N. Evans, *From drill to doctrine: forging the British Army's tactics 1897-1909,* PhD Thesis, King's College, London, 2007.

N. Gardner, *Command and the British Expeditionary Force in 1914*, PhD Thesis, University of Calgary, 2000.

H. Moon, *The Invasion of the United Kingdom: Public Controversy and Official Planning 1888-1918,* PhD thesis, London University, 1968.

R.R. Seim, *Forging the Rapier among Scythes: Lieutenant-General Sir Horace Smith-Dorrien and the Aldershot Command 1907-1912,* MA Thesis, Rice University, Houston, Texas, 1980.

J.D. Sisemore, *The Russo-Japanese War, Lessons Not Learned,* MA Thesis, Faculty of the US Army Command and General Staff College, Fort Leavenworth, Kansas, 2003.

A. Thornton, *The Territorial Force in Staffordshire, 1908-1915,* M Phil Thesis, Birmingham University, 2004.

A. Winrow, *The British Regular Mounted Infantry 1880-1913: Cavalry of Poverty or Victorian Paradigm?* DPhil thesis, University of Buckingham, 2014.

## PUBLISHED SOURCES

M. Adkin, *The Western Front Companion*, London, Aurum, 2013.

Major-General E.A. Altham, *The Principles of War*, London, Macmillan, 1914.

Lord Anglesey, *A History of British Cavalry 1816-1919 Volume VII: The Curragh Incident and the Western Front 1914*, Barnsley, Pen & Sword, 1996.

E. Ashmead-Bartlett, *With the Turks in Thrace*, New York, George H. Doran, 1913.

T. Ashworth, *Trench Warfare 1914-1918: The Live and Let Live System*, London, Pan, 2000.

S. Badsey, *Doctrine and Reform in the British Cavalry 1880-1918*, Aldershot, Ashgate, 2008.

H Bailes,'Technology and Tactics in the British Army, 1866-1900' in R. Haycock, K. Neilson, *Men, Machines and War*, Ontario, Wilfrid Laurier University Press, 1988.

T. A. Baker, *Clacton-on-Sea in old picture postcards volume 1*, Zaltbommel, European Library, 1984.

C.R. Ballard, *Smith-Dorrien*, London, Constable & Company, 1931.

General Sir G. de S. Barrow, *The Fire of Life*, London, Hutchinson & Co., 1941.

S. Batten, 'A school for the leaders: What did the British Army learn from the 1912 army manoeuvres?' *Journal of the Society for Army Historical Research* vol. XCIII no. 373, 2015.

S. Batten, 'The Rivals: Haig and Grierson', *The Douglas Haig Fellowship Records Number 19, June 2017.*

J. Baynes, *Far From a Donkey: The Life of General Sir Ivor Maxse*, London, Brassey's, 1995.

I.F.W. Beckett, *Johnnie Gough, V.C.: A Biography of Brigadier-General Sir John Edmond Gough*, London, Tom Donovan, 1998.

I.F.W. Beckett, *Territorials: A Century of Service*, Plymouth, DRA Printing, 2008.

I.F.W. Beckett and K. Simpson, *A Nation in Arms: A Social Study of the British Army in the First World War*, Manchester, Manchester University Press, 1985.

I.F.W. Beckett and S.J. Corvi, *Haig's Generals*, Barnsley, Pen & Sword, 2006.

Baron Beyens, trans. P.V. Cohn, *Germany before the War*, London, Thomas Nelson and Sons, 1916.

S. Bidwell and D. Graham, *Firepower: British Army Weapons and Theories of War 1904-1945*, London, George Allen & Unwin, 1982.

B. Bond and N. Cave (eds.), *Haig: A Reappraisal 80 years on, 1902-1914*, Barnsley, Pen & Sword, 2009.

B. Bond et al, *Look to your Front – Studies in the First World War by the British Commission for Military History*, Staplehurst, Spellmount, 1999.

T. Bowman and M. Connelly, *The Edwardian Army: Recruiting, Training, and Deploying the British Army, 1902-1914*, Oxford, Oxford University Press, 2012.

C. Brice, *The Thinking Man's Soldier: The Life and Career of Sir Henry Brackenbury 1837-1914*, Solihull, Helion, 2013.

Lieutenant-General Sir T. Bridges, *Alarms and Excursions – Reminiscences of a Soldier*, London, Longmans Green and Co., 1938.

E. Dorn Brose, *The Kaiser's Army: The Politics of Military Technology in Germany during the Machine Age, 1870-1918*, Oxford, Oxford University Press, 2001.

A. Bucholz, *Moltke, Schlieffen and Prussian War Planning*, Providence R.I., Berg Publishers, 1991.

S. Bull, *Trench: A History of Trench Warfare on the Western Front*, Oxford, Osprey, 2010.

S. Bull, *Trench Warfare: A History of Trench Warfare on the Western Front*, London, Compendium, 2003.

C. Byrne, *The Harp and Crown, the History of the 5th (Royal Irish) Lancers, 1902 – 1922*, Lulu Books, 2007.

M. Calthorpe, *Royal Cambridge*, Cambridge, Cambridge City Council, 1977.

Major-General Sir C.E. Callwell: *Field-Marshal Sir Henry Wilson, His Life and Diaries*, London, Cassell, 1927.

A.R. Capon, *His Faults Lie Gently – The Incredible Sam Hughes*, Lindsay Ontario, Floyd W. Hall, 1969.

G.H. Cassar, *The French and the Dardanelles: A study of failure in the conduct of war* (London, George Allen & Unwin, 1971.

Brigadier-General J. Charteris, *At G.H.Q.*, London, Cassell, 1931.

Brigadier-General J. Charteris, *Field-Marshal Earl Haig*, London, Cassell, 1929.

P. Chasseaud and P. Doyle, *Grasping Gallipoli: Terrain, Maps and Failure at the Dardanelles, 1915*, Staplehurst, Spellmount, 2005.

D. Clarke, *World War I Artillery Tactics*, Oxford, Osprey, 2014.

I.F. Clarke, *Voices Prophesying War: Future Wars 1763-3749*, Oxford, Oxford University Press, 1992.

C. von Clausewitz, eds M Howard & P Paret, *On War*, Princeton, Princeton University Press, 1993.

B. Clouting, *Teenage Tommy – Memoirs of a Cavalryman in the First World War*, edited by Richard van Emden, Barnsley, Pen & Sword, 2013.

Sir Arthur Conan Doyle, *The Great Boer War*, London, Smith, Elder & Co., 1900.

D. Cooper, *Haig*, London, Faber and Faber, 1935.

T.S. Crawford: *Wiltshire and the Great War: Training the Empire's Soldiers*, Marlborough, Crowood Press, 2012.

R. Crawley, *Climax at Gallipoli: The Failure of the August Offensive*, Norman, Oklahoma, University of Oklahoma Press, 2014.

P. Crowley, *Loyal to Empire: The Life of Sir Charles Monro 1860-1929*, Stroud, The History Press, 2016.

A. Cruise, *Louis Botha's War: The Campaign in German South-West Africa, 1914-1915*, Cape Town, Zebra Press, 2015.

G.R.S. Darroch, *Deeds of a Great Railway*, London, John Murray, 1920.
General Sir B. de Lisle, *Reminiscences of Sport & War*, London, Eyre & Spottiswoode, 1939.
P. Dennis, *The Territorial Army 1907-1940* (Royal Historical Society Studies in History 51), Woodbridge, Suffolk, Boydell Press, 1987.
R. A. Doughty, 'French Strategy in 1914: Joffre's Own', *The Journal of Military History, Vol. 67, No. 2,* April 2003.
P. Doyle, *Gallipoli 1915,* Staplehurst, Spellmount, 2011.
Colonel J.K. Dunlop, The *Development of the British Army 1899-1914,* London, Methuen, 1938.
Captain J.C. Dunn, *The War the Infantry Knew 1914-1919,* London, Jane's, 1987.
Brigadier-General Sir J.E. Edmonds, *History of the Great War: Military Operations France and Belgium 1914,* London, Macmillan, 1922.
E.J. Erickson, *Gallipoli: Command Under Fire,* Oxford, Osprey, 2015.
N. Evans, 'The British Army and communications, 1899-1914', *Journal of the Society for Army Historical Research,* vol. XCIV no. 379, Autumn 2016.
D. Filsell, 'Thompson Capper and the Forward Slopes', *The Douglas Haig Fellowship Records Number* 18, June 2016.
M. Foley, *Essex at War from Old Photographs*, Stroud, Amberley, 2012.
R. Foley, 'Dumb donkeys or cunning foxes? Learning in the British and German armies during the Great War', *International Affairs 90: 2,* 2014.
R. Foley, 'Easy Target or Invincible Enemy? German Intelligence Assessments of France before the Great War', *The Journal of Intelligence History. Vol. 5, No. 2,* Winter 2005.
G.T. Forestier-Walker, *Umpiring,* Aldershot Military Society Lecture (Monday, April 10, 1911).
Major the Hon. G. French, *The Life of Field-Marshal Sir John French, First Earl of Ypres,* London, Cassell, 1931.
Field Marshal Viscount French of Ypres, *1914,* London, Constable, 1919.
B. Gardner, *Allenby,* London, Cassell, 1965.
N. Gardner, *Trial by Fire: Command and the British Expeditionary Force in 1914,* London, Greenwood, 2003.
N. Gardner, 'The Beginning of the Learning Curve: British Officers and the Advent of Trench Warfare, September-October 1914', *Working Papers in Military & International History, Number 1*, April 2003.
N. Gardner, 'Command and Control in the Great Retreat of 1914: The Disintegration of the British Cavalry Division', *The Journal of Military History (Society for Military History) 63 (1),* 1999.
A. Gilbert, *Challenge of Battle: The British Army's Baptism of Fire in the First World War,* Oxford, Osprey, 2014.
J. Glanfield, *The Devil's Chariots: The origins and secret battles of tanks in the First World War*, Oxford, Osprey, 2013.
General Sir Alexander Godley, *Life of an Irish Soldier*, New York, EP Dutton, 1939.

General Sir Hubert Gough, *Soldiering On: Being the memoirs of General Sir Hubert Gough*, London, Arthur Barker, 1954.

D. Graham, *Against Odds. Reflections on the Experiences of the British Army, 1914-1945*, Basingstoke, Macmillan Press, 1999.

A. Green, *Writing the Great War – Sir James Edmonds and the Official Histories 1915-1948*, London, Frank Cass, 2003.

E. Greenhalgh, *Foch in Command: The Forging of a First World War General,* Cambridge, Cambridge University Press, 2014.

E. Greenhalgh, *The French Army and the First World War,* Cambridge, Cambridge University Press, 2014.

E. Greenhalgh, *Victory through Coalition: Britain and France during the First World War,* Cambridge, Cambridge University Press, 2009.

E. Greenhalgh, 'Did Sir Douglas Haig panic at Landrecies in August 1914 during the Great Retreat?' *Journal of the Society for Army Historical Research* 2013 vol. XCI no. 367.

Lieutenant-General Sir J. Grierson, 'Military Manoeuvres', *Encyclopaedia Britannica* XVII, Cambridge, Cambridge University Press, 1911.

P. Griffith, *Battle Tactics of the Western Front: The British Army's Art of Attack 1916-1918*, London, Yale University Press, 1994.

S.T. Grimes, *Strategy and War Planning in the British Navy, 1887-1918,* Woodbridge, Suffolk, Boydell Press, 2012.

Captain R. Grimshaw, *Indian Cavalry Officer 1914-1915,* Tunbridge Wells, Costello, 1986.

B. Gudmondsson, *The British Expeditionary Force 1914-1915,* Oxford, Osprey, 2005.

Lt.-Col. E. Gunter, 'The von Löbel Annual Reports on the Changes and Progress in Military Matters in 1905', *Royal United Services Institution Journal*, Vol. 49, Issue 333, 1905.

D. Haig (edited by Douglas Scott), *The Preparatory Prologue: Diaries and Letters 1861-1914,* Barnsley, Pen & Sword, 2006.

D. Haig (edited by G. Sheffield and J. Bourne), *War Diaries and Letters 1914-1918,* London, Phoenix, 2006.

General Sir A. Haldane, *A Soldier's Saga*, Edinburgh, William Blackwood, 1948.

Lieutenant-General Sir A. Haldane, *A Brigade of the Old Army 1914*, London, Edward Arnold, 1920.

R.C. Hall, *The Balkan Wars 1912-1913: Prelude to the First World War,* London, Routledge, 2000.

W.K. Hancock and Jean van der Poel (eds), *Selections from the Smuts Papers Volume III, June 1910 – November 1918,* Cambridge, Cambridge University Press, 1966.

Francis Hanford, First Landing, *The story of the first arrival of aircraft on the Halton Estate*, Halton, Trenchard Museum, 2014.

J.P. Harris, *Douglas Haig and the First World War,* Cambridge, Cambridge University Press, 2008.

J.P. Harris, *The Men Who Planned the War: A Study of the Staff of the British Army on the Western Front, 1914-1918*, London, Routledge, 2016.

P. Hart, *Fire and Movement – The British Expeditionary Force and the Campaign of 1914*, Oxford, Oxford University Press, 2015.

J. Hevia, *The Imperial Security State: British Colonial Knowledge and Empire-Building in Asia*, Cambridge, Cambridge University Press, 2012.

R. Holmes, *The Little Field-Marshal: A Life of Sir John French*, London, Jonathan Cape, 1981.

R. Holmes, *Riding the Retreat: Mons to the Marne 1914 Revisited*, London, Jonathan Cape, 1995.

M. Howard, 'Men Against Fire: The Doctrine of the Offensive in 1914' in *Makers of Modern Strategy from Machiavelli to the Nuclear Age* edited by Peter Paret, Oxford, Clarendon Press, 1986.

M. Howard, 'Men Against Fire: Expectations of War in 1914' in *Military Strategy and the Origins of the First World War* edited by Steven E Miller, Princeton University Press, 1985.

General V. Huguet, *Britain and the War, a French Indictment*, London, Cassell, 1928.

J. Hussey, 'Without an Army, and Without any Preparation to Equip One': The Financial and Industrial Background to 1914, *British Army Review* Number 109, April 1995.

J. Hussey, 'Appointing the Staff College Commandant 1906: A Case of Trickery, Negligence, or Due Consideration?' *British Army Review* Number 114, December 1996.

N. Jacobs, *Clacton Past with Holland-on-Sea & Jaywick*, Chichester, Phillimore, 2002.

L. James, *Imperial Warrior: The Life and Times of Field-Marshal Viscount Allenby 1861-1936*, London, Weidenfeld and Nicolson, 1993.

K. Jeffrey, *Field Marshal Sir Henry Wilson: A Political Soldier*, Oxford, Oxford University Press, 2010.

S. Jones, 'Scouting for Soldiers: Reconnaissance and the British Cavalry, 1899-1914', *War in History, Volume 18, no. 4,* November 2011.

S. Jones, *From Boer War to World War: Tactical Reform of the British Army, 1902-1914*, Norman, University of Oklahoma Press, 2012.

S. Jones (ed.), *Stemming the Tide: Officers and Leadership in the British Expeditionary Force 1914,* Solihull, Helion, 2013.

S. Jones, 'Shooting Power: A Study of the Effectiveness of Boer and British Rifle Fire, 1899-1914', *British Journal for Military History, Volume 1, Issue 1,* October 2014.

S. Jones (ed.), *Courage Without Glory: The British Army on the Western Front 1915,* Solihull, Helion, 2015.

P.M. Kennedy, *The War Plans of the Great Powers, 1880-1914*, London, George Allen & Unwin, 1979.

D. Kenyon, *Horsemen in No Man's Land: British Cavalry and Trench Warfare,* Barnsley, Pen & Sword, 2011.

N.A. Lambert, *Sir John Fisher's Naval Revolution,* Columbia, University of South Carolina Press, 2002.

H. Langlois, translated by Captain C.F. Atkinson, *The British Army in a European War*, London, Hugh Rees Ltd., 1910.

A. Leask, *Putty – From Tel-el-Kebir to Cambrai: The Life and Letters of Lieutenant General Sir William Pulteney 1861-1941,* Solihull, Helion, 2015.

D.M. Leeson, 'Playing at War: The British Military Manoeuvres of 1898', *War in History 15,* 2008.

B. Liddell Hart, *History of the First World War*, London, Pan, 1972.

D. Lomas, *Mons 1914, The BEF's Tactical Triumph*, Oxford, Osprey, 2009.

J. Lucy, *There's a Devil in the Drum*, London, Faber & Faber, 1938.

D.S. Macdiarmid, *The Life of Lieut. General Sir James Moncrieff Grierson,* London, Constable & Company, 1923.

Lyn Macdonald, *1914*, London, Penguin, 1989.

S. Marble, *British Artillery on the Western Front in the First World War*, Farnham, Ashgate, 2013.

A. Mallinson, *1914: Fight the Good Fight: Britain, the Army and the Coming of the First World War,* Ealing, Bantam Press, 2013.

J. Marshall-Cornwall, *Haig as Military Commander*, London, Batsford, 1973.

A.M. Matin, 'The Creativity of War Planners: Armed Forces Professionals and the pre-1914 British Invasion-Scare Literature', *English Literary History 78*, 2011.

Col. H.A.R. May, *Memories of the Artists Rifles*, London, Howlett & Son, 1929.

H.B. McCartney, *Liverpool Territorials in the First World War*, Cambridge, Cambridge University Press, 2005.

Frank R. McCoy, Notes on the German Maneuvers, *Journal of the U.S. Cavalry Association 14* (July 1903).

G. Mead, *The Good Soldier, The Biography of Douglas Haig,* London, Atlantic, 2007.

C. Messenger, *Call to Arms: The British Army 1914-1918*, London, Weidenfeld & Nicolson, 2005.

K.W. Mitchinson, *England's Last Hope: The Territorial Force 1908-1914,* Basingstoke, Palgrave Macmillan, 2005.

K.W. Mitchinson, *The Territorial Force at War, 1914-1916,* Basingstoke, Palgrave Macmillan, 2014.

D. G. Morgan-Owen, 'Cooked up in the Dinner Hour? Sir Arthur Wilson's War Plan, Reconsidered', *The English Historical Review,* Vol. CXXX, No. 545, 2015.

A.J.A. Morris (ed.), *The Letters of Lieutenant-Colonel Charles a Court Repington*, Stroud, Sutton Publishing, 1999.

A.J.A. Morris, *Reporting the First World War: Charles Repington, The Times and the Great War*, Cambridge, Cambridge University Press, 2015.

N. Murray, *The Rocky Road to the Great War: The Evolution of Trench Warfare to 1914*, Washington DC, Potomac Books, 2013.

R. Neillands, *The Old Contemptibles: The British Expeditionary Force, 1914*, London, John Murray, 2004.

Brevet Colonel G.H. Ovens, 'Fighting in Enclosed Country with some Notes from the Essex Manœuvres', *Royal United Services Institution Journal Volume 49, Issue 327*, 1905.

David Owen, *The Hidden Perspective: the Military Conversations, 1906-1914*, London, Haus Publishing, 2014.

P. Parker, *The Old Lie, The Great War and the Public-School Ethos*, London, Constable, 1987.

G. Phillips, 'The Obsolescence of the Arme Blanche and Technological Determinism in British Military History', *War in History* IX (2002).

W. Philpott, *Attrition: Fighting the First World War*, London, Abacus, 2014.

D. Porch, *The March to the Marne: The French Army 1871-1914*, Cambridge, Cambridge University Press, 1981.

E.A. Pratt, *The Rise of Rail-power in War and Conquest 1833-1914*, London, P.S. King & Son, 1915.

E.A. Pratt, 'The Role of Railways in the War' in *The Great War, The Standard History of the All-Europe Conflict Volume 6*, London, Amalgamated Press, 1916.

R. Prior, *Gallipoli, The End of the Myth*, New Haven, Yale University Press, 2009.

R. Prior and T. Wilson, *Command on the Western Front: The Military Career of Sir Henry Rawlinson 1914-1918*, Oxford, Blackwell, 1992.

Sir W. Raleigh, *The War in the Air Volume I*, Oxford, Clarendon Press, 1922.

A. Rawson, *The British Army 1914-1918*, Staplehurst, Spellmount, 2006.

W. Reid, *Douglas Haig: Architect of Victory*, Edinburgh, Birlinn, 2009.

Revue militaire des armées étrangères No. 1028 (July 1913), Paris, Librairie Chapelot, 1913.

M. Richardson, 1914: *Voices from the Battlefields*, Barnsley, Pen & Sword, 2013.

S. Robbins, *British Generalship during the Great War – The Military Career of Sir Henry Horne*, Farnham, Ashgate Publishing, 2010.

S. Robbins, *British Generalship on the Western Front 1914-1918. Defeat into Victory*, London, Frank Cass, 2005.

D. Russell, *Winston Churchill: Soldier: The Military Life of a Gentleman at War*, London, Brassey's, 2005.

W.M. Ryan, *Lieutenant Colonel Charles a Court Repington. A Study in the Interaction of Personality, the Press, and Power*, London, Garland Publishing, 1987.

M. Samuels, *Command or Control?: Command, Training and Tactics in the British and German Armies, 1888-1918*, London, Frank Cass, 1995.

A. Saunders, *Trench Warfare 1850-1950*, Barnsley, Pen & Sword, 2010.

B. Scott, *Galloper Jack*, London, Macmillan, 2003.

G.K. Scott-Moncrieff, 'Land for Military Training. An Appeal and a Suggestion,' *Blackwood's Magazine CLXXX* (August 1906), pp.153-163.

A. Scrimgeour, *Scrimgeour's Small Scribbling Diary 1914-1916*, London, Conway, 2008.

J.E.B. Seely, *Fear and Be Slain: Adventures by Land, Sea and Air*, London, Hodder and Stoughton, 1931.

M. S. Seligmann, *Spies in Uniform: British Military and Naval Intelligence on the Eve of the First World War,* Oxford, Oxford University Press, 2006.

A. Seldon and D. Walsh, *Public Schools and the Great War,* Barnsley, Pen & Sword, 2013.

G. Sheffield, *Douglas Haig: From the Somme to Victory*, London, Aurum, 2016.

G. Sheffield, *Forgotten Victory – The First World War: Myths and Realities*, London, Headline, 2001.

G. Sheffield and D. Todman (editors), *Command and Control of the Western Front: The British Army's Experience 1914-18,* Staplehurst, Spellmount, 2007.

G. Sheffield, 'The Somme: A Terrible Learning Curve', *BBC History Magazine*, July 2011.

P. Simkins, *Kitchener's Army: The Raising of the New Armies 1914-1916*, Barnsley, Pen & Sword, 2007.

Andy Simpson, *Directing Operations: British Corps Command on the Western Front 1914-1918,* Staplehurst, Spellmount, 2006.

A.J. Smithers, *The Man who Disobeyed, Sir Horace Smith-Dorrien and his Enemies*, London, Leo Cooper, 1970.

General T. Snow, edited by Dan Snow and Mark Pottle, *The Confusion of Command: The War Memoirs of Lieutenant General Sir Thomas D'Oyly Snow, 1914-1915*, Barnsley, Frontline Books, 2011.

L. Sondhaus, *Franz Conrad von Hötzendorf: Architect of the Apocalypse*, Boston, Brill Publishers, 2000.

E.L. Spears, *Liaison 1914*, London, William Heinemann, 1930.

E.M. Spiers, *Haldane: An Army Reformer*, Edinburgh, Edinburgh University Press, 1980.

E.M. Spiers, 'The British Cavalry 1902-1914', *Journal of the Society for Army Historical Research, vol. LVII no. 230*, Summer 1979.

E.M Spiers, 'Rearming the Edwardian Artillery', *Journal of the Society for Army Historical Research, vol. LVII no. 231,* Autumn 1979

E.M. Spiers, 'Reforming the Infantry of the Line 1900-1914', *Journal of the Society for Army Historical Research*, vol. LIX no. 238, Summer 1981.

J. Sunderland and M. Webb, *All the Business of War: the British Army Exercises of 1913, the British Expeditionary Force and the Great War*, Newbury, Press to Print, 2013.

P. Takle, *The Affair at Nery*, Barnsley, Pen & Sword, 2006.

J. Terraine, *Douglas Haig, The Educated Soldier,* London, Hutchinson, 1963.

J. Terraine, *The Smoke and the Fire, Myths and Anti-myths of War 1861-1945*, London, Book Club Associates, 1980.

J. Terraine, *White Heat*, London, Sidgwick and Jackson, 1982.

W. Thompson, *Morland – Great War Corps Commander: War Diaries & Letters, 1914-1918,* Leicester, Troubadour, 2015.

J. Tomes, *Balfour and Foreign Policy: The International Thought of a Conservative Statesman,* Cambridge, Cambridge University Press, 1997.

Captain A.H. Trapmann, *The Greeks Triumphant,* London, Forster Groom & Co., 1915.
T. Travers, 'The Offensive and the Problem of Innovation in British Military Thought 1870-1915', *Journal of Contemporary History Vol. 13*, 1978.
T. Travers, 'The Hidden Army: Structural Problems in the British Officer Corps, 1900-1918', *Journal of Contemporary History Vol. 17,* 1982.
T. Travers, *The Killing Ground. The British Army, the Western Front and the Emergence of Modern Warfare 1900-1918*, London, Allen & Unwin, 1987.
D. van der Meulen, *Koning Willem III,* Amsterdam, Boom Publishers, 2013.
R. van Emden, *The Soldier's War: The Great War Through Veterans' Eyes*, London, Bloomsbury, 2008.
H.P. van Tuyll van Serooskerken, *Small Countries in a Big Power World: The Belgian-Dutch Conflict at Versailles, 1919,* Leiden, Brill Books, 2007.
Major-General J. Vaughan, *Cavalry and Sporting Memories,* Bala, Bala Press, 1954.
K. Walker, *The History of Clacton*, Clacton-on-Sea, A. Quick & Co., 1966.
Philip Warner, *Field Marshal Earl Haig*, London, Bodley Head, 1991.
R. Wellbye, *Road Touring in Eastern England*, London, Larby, 1921.
G. Whale, *British Airships: Past, Present and Future*, London, John Lane, 1919.
Major A.C. Whitehorne, *The History of the Welch Regiment,* Cardiff, Western Mail and Echo, 1932.
Andrew Whitmarsh, 'British Army Manoeuvres and the Development of Military Aviation, 1910-1913', *War in History 14 (3),* 2007.
A.Wiest, *Haig: The Evolution of a Commander,* Dulles, Virginia, Potomac, 2005.
G. Winton, *Theirs Not To Reason Why – Horsing the British Army 1875-1925*, Solihull, Helion, 2013.
C. Wolmar, *Engines of War, How Wars Were Won and Lost on the Railways*, London, Atlantic, 2012.
P.F. Wright, *The Royal Flying Corps in Oxfordshire 1912-1918*, Leamington Spa, Warwick Printing, 2015.
D. Zabecki, 'The Dress Rehearsal: Lost Artillery Lessons of the 1912-1913 Balkan Wars', *Field Artillery,* August 1988.
Count R. von Zedlitz-Trützschler, *Twelve Years at the Imperial German Court*, London, Nisbet & Co., 1924.

## WEBSITES

John Birch, Terrible air fatality at the Manoeuvres, http://www.hertsmemories.org.uk/content/herts-history (accessed 30/8/2015).
Canadian Officers Notes on 1912 Manoeuvres part 1 https://canadaatwarblog.wordpress.com/2015/06/06/canadian-officers-notes-on-british-army-manoeuvres-sept-16th-19th-1912-part-i/ (accessed 30/8/2015).

Canadian Officers Notes on 1912 Manoeuvres part 2 https://canadaatwarblog.wordpress.com/2015/06/07/canadian-officers-notes-on-british-army-manoeuvres-sept-16th-19th-1912-part-ii/ (accessed 30/8/2015).

Canadian Officers Notes on 1912 French Manoeuvres https://canadaatwarblog.wordpress.com/2015/06/07/canadian-officers-notes-on-french-army-manoeuvres-sept-11th-17th-1912/ (accessed 30/8/2015).

D. G. Morgan-Owen, *Lessons Learned from Gallipoli,* http://defenceindepth.co/2015/05/11/lessons-learned-from-gallipoli (accessed 30/8/2015).

W. Philpott, 'Beyond the Learning Curve: The British Army's Military Transformation in the First World War', RUSI Website, 10 November 2009. https://rusi.org/commentary/beyond-learning-curve-british-armys-military-transformation-first-world-war (accessed 30/8/2015).

Fanny Wale, *A record of Shelford Parva* (1920), https:/www.littleshelfordhistory.co.uk. (accessed 30/8/2015).

*The Warwick Bioscope Chronicle*: His Majesty's Manoeuvres, East Anglian Film Archive, http://www.eafa.org.uk/catalogue/543 (accessed 30/8/2015).

Rev. W. Burnard Wilder, Notes, http://www.great bradley.suffolk.gov.uk/History/wilderfamily2.htm (accessed 30/8/2015).

# Index

## INDEX OF PEOPLE

**INDEX OF PLACES**

## INDEX OF MILITARY FORMATIONS & UNITS

## INDEX OF GENERAL & MISCELLANEOUS TERMS